Software Testing

For

Financial Services Firms

By

Hal McIntyre
Megan Johnson
Deborah Fortuna

Text and art copyright © 2003 by H. McIntyre, M. Johnson, D. Fortuna. All rights reserved. No part of this book may be reproduced or transmitted in any form, by any means without the prior written permission of any of the authors.

ISBN: 0-9669178-4-7

Printed in the United States of America

Published and distributed by The Summit Group Press, a division of The Summit Group.

For general information on The Summit Group Press or for permission to photocopy items for corporate, personal or educational use, please call our Wall Street office at (212) 328-2500.

Limit of Liability/Disclaimer of Warranty: The authors and publisher have used their best efforts in preparing this book. The Summit Group, The Summit Group Press and the authors make no representations or warranties with respect to the accuracy or completeness of the contents of this book and specifically disclaim any implied warranties of merchantability or fitness for any particular purpose. Neither The Summit Group, The Summit Group Press nor the authors shall be liable in any event for any loss of profit or any other damages, including but not limited to special, incidental, consequential, or other damages.

Cover photo and design: Lee Titone

Forward

The authors would like to thank the staff of The Summit Group and Securities Operations Forum for their support, and would especially like to recognize the continuous editing and feedback provided by Kim McIntyre, Eileen Haeger, Scott Porter, Kelly McIntyre and Ari Sternberg.

The authors would also like to thank the following vendors for providing much of the information about their products that is presented in this book:

- AutoTester
- Compuware
- Empirix
- Mercury Interactive
- McCabe
- Radview
- Rational
- Segue

The contributions of all of these people were invaluable in producing a book of this scope. Any errors are the authors'.

June 15, 2003

Hal McIntyre
Managing Partner
hal.mcintyre@tsgc.com

Megan Johnson
Senior Associate
megan.johnson@tsgc.com

Deborah Fortuna
Senior Associate
deb.fortuna@tsgc.com

The Summit Group
48 Wall Street / 4th Floor
New York, NY 10005
(212) 328-2500
www.tsgc.com

Contents

I. Introduction 1
   Systems Development Life Cycle 1
      Software Development 2
      Implementation 3
      Production 3
   Testing 3
      Defect Resolution 4
      Testing vs. Quality Assurance 5
   Organization of the Book 6
   Summary 6

II. **Key Players in the Testing Process** 9
   The Project Management Team 9
      Project Manager 9
      Project Sponsor 9
      Development Manager 10
      Test Manager 10
      End User Representative(s) 10
      Project Administrator 10
      Systems Manager 10
      Auditor 10
   The Development Team 11
      Business Analyst(s) 11
      Database Administrator 11
      Developers 11
      Testers 12
   The Test Team 12
      Test Section Managers 12
      Test Team Administrator 13
      Testers 13
      Test Tools Implementer 13
      Tool Administrator 14
   The End Users 14
      User Experts 14
      Testers 14
   The Testing Role 15
   Team Support Roles 16
   Summary 17

III. **Test Types** 19
   Requirements Reviews 19
      Business Review 19
      Technology Design Review 19
      Data Design Review 20
   Testing Performed by the Development Team 20
      Unit Tests 20
      System Integration Test 21
      Systems Technical Tests 23
   Testing Performed by the Test Team 24
      Business Technical Tests 24

    Business Functional Tests 24
    Business Integration Test 25
    Regression Tests 26
  Testing Performed by the End Users 26
    User Acceptance Test 27
    Parallel Test 28
    Conversion Test 29
  Special Testing Situations 29
    Comparison Testing 29
    Prototype Testing and Pilot Testing 30
  Summary 30

IV. Test Project Categories 33
  Understanding the Business 33
  Understanding Interfaces with Existing Businesses 34
  Testing New Applications (Categories 1 and 2) 34
    Category 1 - Testing a New Application from a Vendor 34
    Category 2 - Testing a New Internally Developed Application 39
  Testing Changes to an Existing System 42
    Category 3 - Testing Vendor Maintained Changes 42
    Category 4 - Testing Internally Maintained Applications 44
  Category 5 - Conversion Testing 46
    Data Conversions 47
    Source Documentation 47
    Conversion Processing 48
    Test Types 48
  Summary 49

V. Getting Started 51
  Selecting the Testing Methodology 52
    Quality Management 53
  Project Management 53
  Test Management Conventions 54
  Standards Organizations 55
    Institute of Electrical and Electronics Engineers, Inc., (IEEE) – Software Engineering Standards Committee (SESC) 56
    The International Organization for Standards (ISO) 57
    Project Management Institute (PMI) 57
    Software Engineering Institute – Capability Maturity Model 58
  Documentation Standards 59
  Software Testing Best Practices 60
    Best Practices - Basics 61
    Best Practices - Foundational 62
    Best Practices - Incremental 63
  Sizing the Project 64
  Summary 66

VI. Master Test Plan 67
  Master Test Plan Objectives 67
    Document the Test Scope 67
    Provide Direction to the Test Team 67
    Provide Information to Interested Parties 67
  Master Test Plan Standards 67

    Master Test Plan Overview 68
        Summary Test Plan 68
        Detailed Test Plans 68
        Project Plan 69
    Summary Test Plan Contents 69
    Establishing the Summary Test Plan 70
        Definition 70
        Assumptions 71
        Test Scope 73
        Dependency Projects 77
        Test Phases 78
        Test Calendar 80
        Developing the Timeline 83
        Adjusting the Timeline 83
        Databases and Environment Plan 86
        Human Resource Plan 89
        Automation Plan 92
        Quality Assurance Plan 92
        Other Resources Plan 92
        Risk Management Plan 92
        Communications Plan 92
        Budget Plan 93
        Reviews and Approvals Plan 93
        Project Plan 93
    Summary 94

VII. Detailed Test Plans 95

    Establishing Detailed Test Plans 96
        Introduction 96
        Scope 96
        Test Calendar 99
        Test Development 99
        Execution Checklist 101
        Promotion Requirements 102
    Project Plan 102
    Summary 103

VIII. Test Project Plan 105

    Project Plan Contents 105
    Test Project Plan Reporting 107
        By Milestone 107
        In-Progress Tasks 107
        Late Tasks 107
        Tasks to Begin During a Specified Time 108
        By Resource 108
    Test Project Plan 108
        Test Calendar 108
        Other Testing-Related Tasks 108
    Summary 109

IX. Business Functional Testing Process 111

    Test Preparation 111
        Data Sources 111

    Business Scenarios 113
    Test Case 113
    Test Procedure 114
    Business Scenario Documentation 116
    Test Case Documentation 116
    Test Case Documentation 118
    Test Procedure Documentation 118
  Database/Environment Preparation 121
  Test Execution 122
    Test Transaction Entry 122
    Test Execution Documentation 122
  Test Verification 122
  Issues Tracking 122
    Issues Management 123
    Issues Documentation 124
  Summary 128

X. Business Technical Testing Process 129
  Preparing Business Technical Tests 129
    User Interfaces (UI) 131
    Reports 143
    Files and Databases 146
    Application Interfaces 149
  Database/Environment Preparation 152
  Test Execution 152
  Issues Tracking 152
  Summary 152

XI. Managing Test Project Processes 153
  Task Management 154
  Test Management 154
    Test the Test Process 154
    Execute/Control the Tests 154
    Define Meaningful MIS 156
    Reviewing the Status 159
    Adjust The Plans 160
  Quality Management 164
    Define Meaningful MIS 165
    Report Categories 168
    Report Review Frequency 172
    Adjust the Plans 172
  Summary 177

XII. Managing Test Project Administration 179
  Scope Creep Management 179
    Functional Scope Creep 179
    Test Case Scope Creep 180
    Define Meaningful MIS 180
    Reviewing the MIS 184
    Adjusting the Plans 185
  Financial Management 186
    Develop Meaningful MIS 187
    Reviewing the MIS 192

Reviewing the Plans 193
Risk Management 194
   Develop Meaningful MIS 195
   Reviewing the MIS 198
   Adjusting the Plans 198
Communication Management 199
   Stakeholders 199
   Meetings 199
   Documentation 202
Summary 206

XIII. Managing Testing Automation 209
White Box Testing 209
Black Box Testing 209
Manual Testing 210
   Staffing Requirements 210
   Key Steps 210
   Advantages/Disadvantages of Manual Testing 211
Automated Testing 211
   Potential Automation Areas 211
   Staffing Requirements 212
   Key Steps 212
   Factors Affecting Automated Testing 212
   Benefits of Automated Testing 213
Comparing Manual vs. Automated Testing 214
Tools Supporting Testing During the SDLC 214
   Test Design Tools (Logical Design and Physical Design) 215
   Test Execution and Comparison Tools 215
   Test Management Tools 216
Building Maintainable Tests 217
   Test Libraries 217
   Test Suites 217
Tool Vendors 217
Automated Tool Functions 217
   Input 217
   Validation 217
   Tracking 218
   Scheduler 218
   Detail/Summary Results 218
   Process Management 218
   Service Level Management 218
Tool Evaluation 219
Tool Comparison Criteria 219
Technical Considerations 220
   Testing Methodology 220
   Using Multiple Protocols 220
   Executable 221
   Pricing Model 221
   Java Components 221
   IP Spoofing 221
   General Performance 221
Summary 221

XIV. Closing the Test Project 223
Prior To Implementation 223
- Identify Software to Move into Production 224
- Deferred Functions 225
- Sign-offs 225

Post Implementation 226
- Developing Test Plans for Ongoing Activity 226
- Rewarding/Recognizing People 227
- Closing the Financials 228
- Post Project Review 229

Summary 231
- Pre-Implementation 232
- Post Implementation 232

XV. Automated Test Vendors 233
AutoTester 233
Compuware 235
Mercury Interactive 241
Empirex 246
McCabe 249
Radview 251
Rational 253
Segue 256
Summary 260

XVI. Test Outsourcing 261
Request for Proposal 261
Outsourcing Guidelines 264
Summary 265

Appendix I - Test Team Job Descriptions 267
Test Team Manager 268
- Responsibilities 268
- Skills/Experience 268

Test Section Manager 269
- Responsibilities 269
- Skills/Experience 269

Test Administrator 269
- Responsibilities 269
- Skills/Experience 270

Testers 270
- Responsibilities 270
- Skills/Experience 270

Test Tool Implementer 270
- Responsibilities 270
- Skills/Experience 271

Appendix II - Automation Vendors 272
Auto Tester 272
Compuware 273
Mercury Interactive 274
Empirex 275

McCabe...276
Radview..276
Rational..277
Segue Software...278

Appendix III - Definitions Used to Compare Testing Tools 279
Feature Attribute Definitions 279
Other Applications Support 279
One Script for All Applications 279
Benefit Attribute Definitions 279
GUI Based Applications Testing 279
Scripting Language..279
Script Development Modules 279
Test Creation 279
Predictive Capability 279

Appendix IV - Software Acceptance Criteria 280
Initial Software Delivery 280
Code Repair Timeliness..280
Code Quality...281
Service Level Agreement..281

Appendix V - Case Studies 282
Case Study – Compuware / AZUR-GMF 282
Case Study – Compuware / CIBC 283
Case Study – Compuware / First Horizon Home Loans 286
Case Study – Compuware / Royal & SunAlliance 288
Case Study – Mercury / John Hancock Financial Services 290
Case Study – Mercury / NASDAQ 292
Case Study – Rational / Industrial & Financial Systems 295

Appendix VI - Master Test Plan Checklists 301
Test Management Checklist..301
Master Test Plan Checklist 302
Test Management Planning Checklist 304
Functional Testing Execution Checklist 305

Appendix VII - Glossary 306

Appendix VIII - Index.. 316

Appendix IX - Author Biographies 320

I. INTRODUCTION

Many books have been written on the topic of software testing. Most of these books were written by academics and provide generic solutions so that people in every industry can use them.

This book was written exclusively for professionals in the Financial Services Industry by practitioners who have over 65 years of combined experience in building and testing financial applications. The authors have obtained their experience in a variety of firms while working in software development, testing and operational positions.

The book presents a thorough testing process that can be implemented manually, and also evaluates several automated testing tools that can support the process. There are many practical examples, Guidelines and Checklists for managers who need to test applications of all sizes and complexity; and which can be useful to banks, brokers, investment managers, insurance companies, etc.

There are a number of definitions that are presented throughout the book. Some of these definitions are the same as in other books, some are used by Standards organizations and some are the authors' own. Whether a Test Manager uses these definitions, or creates his/her own, it is important to have everyone involved in the testing effort use the same terms in the same way. This will save considerable time and reduce frustrations. Key terms are capitalized throughout the book to help the reader easily identify testing terminology.

> "Real knowledge is of two kinds. We know a subject ourselves, or we know where we can find information upon it."
> — Samuel Johnson

There are also many different approaches to testing; while any approach can be successful to some degree, a firm can increase its potential for a successful Project by applying the most appropriate solution. Testing is not a one-time activity, and it should be conducted throughout the Project, during all phases of the Systems Development Life Cycle (SDLC). Managers who skip steps risk increasing overall costs.

Systems Development Life Cycle

The SDLC approach[1] was established as a result of the positive and negative experiences of Application Developers when the first computer Applications were Programmed. Project Managers found that if they followed a rigorous series of steps that demanded agreed upon Requirements, Design and Test Plans; they would get a better software product, faster and with less wasted effort.

There are many variations on the SDLC concept. The basic one is called the 'waterfall' approach because each step must be performed and completed before the next step is started. This approach worked well for Applications with large centralized teams of Developers. However, the move to client/server architecture and now to web-based and thin client architectures has forced some changes since most Development involves smaller portions of the Application going through independent parallel and repetitive cycles.

Even within the waterfall methodology, the process is not a series of one-time events. Development requires testing, which leads to Production, which identifies additional Development work that needs to be performed. In order to build high-quality Code, the overall process of building software Applications needs to be

---

[1] This methodology is also called the Project Development Life Cycle (PDLC), the Project Life Cycle (PLC) or the Product Life Cycle (also PLC).

understood, and firms typically customize the generic SDLC approaches to meet their particular needs.

The steps that are important in any SDLC are shown in the following chart. The related testing activities are introduced later in the chapter and discussed in detail in Chapter III.

STAGE OF DEVELOPMENT	SDLC DEFINITION
Concept and Analysis	Projects start with an idea that is defined, analyzed and evaluated at a high level and which has to be approved by some authority before the Project continues.
Requirements	Business Requirements: The concept has to be described in detail by the business people who will use the Application. This description must be written down in as much detail as the End Users can provide.
	Functional Requirements: The Business Requirements Document has to be evaluated by Business Analysts and Systems Analysts who will add more details so that Developers can use the Document as a basis for writing Code after the Design has been completed.
Design	Once the Business and Functional Requirements have been described in detail, technical experts convert that information into a Design that defines the Files required, the grouping of Functions into Modules, the technology that will be used, etc.
	Design is usually conducted first at a high level and subsequently at a lower level of detail.
Programming	Developers only should start to write Code after the Application (or Change) has been fully defined and Designed. Software Engineering Standards often are used to help the Developers produce high quality Code.
Testing	The Testing process will be discussed throughout this book.
Implementation	Applications that pass all of the required tests can be moved into Production. There are many other activities that are involved in moving an Application into Production which are beyond the scope of this book.
Production	Once in Production an Application may need Changes. There should be a formal process for this on-going aspect of Development that is similar to the full Development process.

Figure 1 – The Summit Group's Systems Development Life Cycle

There are three major activities that were introduced in the previous chart:

- Software Development
- Implementation
- Production

Authors' note: Projects that move directly from an idea to Coding can be fun for the Development Team and Test Team but are seldom successful.

These three activities are described in the following sections.

Software Development

Software Development, also called Software Engineering, is a process that requires very extensive training, experience and detailed procedures, and which is beyond the scope of this book. The SDLC process results in computer instructions that are organized into a number of different groupings.

GROUPING	DEFINITION
Code	Code consists of individual lines that contain the Computer instructions
Module	A Module is a group of lines of Code that performs a specific Task
Program	A Program is a group of Modules that work together to perform a specific Function
Application	An Application is a group of Programs that meet a specific business requirement
System	A System is a group of interconnected Applications that meet all of a business's needs

Figure 2 – Software Development Definitions

The definitions in Figure 2 will be used throughout this book.

Implementation

The actual Implementation of an Application is beyond the scope of this book.

Production

Once an Application has met the pre-established criteria for success, it can be moved into Production. Once an Application is in production, Changes are always required. This Maintenance process requires that new Code be written and tested. The process of testing Changes to an Application is included in this book.

The primary focus of this book is testing.

Testing

The objective of testing is to find Defects, track the Developers' repair of the Defect, and verify that the Defects have been eliminated. A Defect is either:

- An Error in the Code that delivers an incorrect response (or no response)
- Correct Code that does not meet the Business Requirements

The major testing activities in a project consist of Planning, Execution, Controlling, and Closing, and there are a number of different Test Types that should be performed at different stages of the SDLC to identify Code Defects as early in the process as possible. These Test Types are defined in Chapter III, and support the various stages of the SDLC as shown in the following chart.

STAGE OF DEVELOPMENT	TEST TYPES
Concept and Analysis	None
Requirements	After the Business and Functional Requirements are completed, they should be subjected to a full Business Review.
Design	The appropriate technical and business experts should evaluate the Design by performing the following reviews: • Technical Design Review • Data Design Review
Programming	There are a series of tests that Developers are expected to perform as they work in order to ensure that they are producing Code that meets the defined Requirements, and which works correctly. These tests are: • Unit Tests • Systems Integration Test • Systems Technical Tests
Testing	After the Developers have completed Coding, a series of additional tests must be performed before the Code is delivered to the End Users. Depending upon the Category of the Testing Project, the Developers, Test Team or End Users may perform one or more of the following tests: • Business Technical Tests • Business Functional Tests • Business Integration Test • Regression Tests Once the Project Managers feel that they have Developed a usable Program, Module or Application, the End Users of the Application are asked to independently test the Application to verify that it meets the Requirements and works correctly. This is called the User Acceptance Test (UAT).
Implementation	Applications that pass the UAT may be subjected to additional Systems Technical Tests, and if successful, are ready to be installed in Production. If the new Application is replacing an existing Application, the End Users may have to perform a Conversion Test and a Data Conversion. Depending upon the Category of the Testing Project, the End Users also may require a final test of the Application before moving it into Production. This final test is usually the Parallel Test. In a Parallel Test, the data that is entered into the Application that is already in Production will also be entered, in parallel, into the new Application and the results of both Applications will be compared.
Production	Once in Production an Application may need Changes and the End Users may identify additional Defects.

	Testing is required to verify that the new Code meets the Requirements and works correctly, and that the new Code does not negatively affect the existing Code. For small Changes, the new Code might only be subjected to a Regression Test, while larger deliveries of Code could require more extensive testing.

Figure 3 – Test Categories

Defect Resolution

The purpose of a testing process is to identify Defects and ensure that the Defects are repaired and re-tested. Many years ago, MIT demonstrated that the later in the development process that a Defect is discovered, the greater the cost to repair it. For every $1.00 that it takes to fix a Defect during the Programming Phase, it takes less to repair it during the Requirements and Design Phases, and significantly more to repair during the later phases. It was this realization that caused firms and IT professionals to adopt structured approaches to Developing Applications such as an SDLC methodology.

SDLC PHASE	COST TO REPAIR
Concept and Analysis	0
Requirements	.10 - .30
Design	.30 - .60
Programming	$1.00
Testing	12.00 - 40.00
Implementation	40.00 - 60.00
Production and Maintenance	30.00 - 100.00

Figure 4 - Relative Cost to Fix Defects

Finding Defects earlier in the SDLC can reduce costs in several ways:

- Since Defects are found continuously throughout the testing process, the cost and effort of managing Defects increases significantly with time. This is because there will be some Defects that are identified and being worked on while more Defects are being discovered, and by the time that the first Defects are ready for Re-test, some new Defects have been found and the second batch is being processed. All of the Defects that are found in a Test Phase are not normally completed at the same time, and some may not be resolved until a later Test Phase. Some Defects are easy to fix and are returned quickly, others need additional information and may take days or weeks. This all leads to increased administration and management cost and effort.

  > **Authors' note:** If you identify Defects earlier in the process, you will spend less on repair and have more money available for Development.

- As the Project moves ahead, the Programmers who wrote the initial Code may no longer be available to work on the repairs. This extends the learning curve as new people are added and introduces the potential for additional Defects.
- And typically, as some of the Code moves into UAT, the End Users identify new Changes that are required because of incomplete or inaccurate Requirements or an evolution of the business needs. This effort will compete with the Defect repair process for resources.

The Yankee Group found that as much as 89% of an average large firm's IT Budget is dedicated to Maintenance, much of which results from Defects that are discovered only after the Code has moved into Production. The solution is to identify Defects as early in the Development process as possible, since firms that do not test adequately, and early, increase the proportion of their Budget that is required to fix Defects. If the Defects are not discovered by Inspections, they should be found by testing.

Testing vs. Quality Assurance

Software Testing is not the same as Quality Assurance. Both are important in order to Implement a high quality Application.

Software Testing

The definition of testing according to the ANSI/IEEE 1059 Standard is that "testing is the process of analyzing a software item to detect the differences between existing and required conditions (i.e., Defects) and to evaluate the features of the software item." This means:

> **Authors' note:** If you don't test under Conditions similar to those that will be encountered during Production, you will not be testing properly.

- Does it meet the Functional Requirements?
- Does it work correctly?

Testing software involves running the software in a controlled Environment, to:

- Verify that the Functional Requirements represent what the user really needs/wants
- Validate that the Code actually does what the Functional Requirements specify
- Identify Code that does not work correctly

Testing minimizes the Risk of product failure by providing some assurance that the Application will work in Production under the Conditions that were tested.

Quality Assurance

Quality Assurance is a process that is used to support the Development of a product or an Application. Quality Assurance in software Development is the process of monitoring all of the activities that are associated with software Development, from Concept through Production. It involves the entire Software Development Life Cycle.

> **Authors' note:**
> Testing = Detection
> QA = Prevention

In a full QA process, the emphasis is on prevention, where testing is focused on detection. Prevention involves:

- Following previously defined procedures
- Implementing Development and Testing Standards
- Generalizing from specific Defects to identify other areas at Risk

Quality Assurance activities often are conducted in parallel with Application Development. Without a QA process in place, Projects run the Risk of one or more of the following Defects:

- Incomplete or inaccurate Requirements
- Slow or inconsistent resolution of Defects and/or Issues
- Inaccurate or incomplete Test Cases
- Incorrect analysis of Test Results
- Unknown or inaccurate Project status
- Too much or too little testing

While this book focuses on testing, several of the activities that are required to prevent Defects are covered.

Organization of the Book

The first chapters provide some background and define the terminology that will be used throughout the book. Chapter II discusses the key participants in the testing process, Chapter III defines the various Test Types that can be performed, and in Chapter IV we review the Test Project Categories that can exist when validating an Application.

Chapter V starts the discussion about Planning, Executing and Controlling Projects. This continues in Chapter VI, which describes how to build a Master Test Plan, and then in Chapter VII, we show how the High Level Plan is transformed into a series of Detailed Test Plans. After the Plans are prepared, the Project needs to be managed, and Chapter VIII describes how all of the pieces can be pulled together to establish a full Project Plan.

In Chapter IX, we discuss the Testing Process for Business Functions, and in Chapter X we identify some of the common Business Technical Testing Conditions that are seen most often in Testing Projects. Managing this Test Process is described in Chapter XI, and managing the related Administrative Processes is reviewed in Chapter XII. Chapter XIII reviews the important aspects required for Managing Testing Automation and Chapter XIV identifies the steps that a Project Manager should take to close a testing Project.

Chapter XV evaluates a number of different testing vendors and their tools, and Chapter XVI discusses how a Project Manager can determine if the testing effort can be outsourced, and if so, how this process can be managed.

And, we have also assembled several Appendices that provide supplementary information about the Testing Process, as well as some specific Case Studies and Checklists to help start the testing effort.

Summary

This chapter covered the following key points:

- A testing effort cannot be successful if the Application was not programmed under some Systems Development Life Cycle methodology. It is almost impossible to test effectively without pre-defined Requirements.
- A complete Test Process involves a number of different testing activities that parallel the steps in an SDLC.
- The earlier in the SDLC that testing begins, the lower the overall cost.
- Quality Assurance is based upon prevention while testing is based upon detection.

Guiding Themes

There are three guiding themes that are used throughout this book:

1. Inspect, Don't Expect
2. Plan, Execute, Control and Close
3. Measure, Monitor and Manage

Guiding Principles

The following points are the Guiding Principles that should be used for any testing Project.

1. A comprehensive Test Plan should include all of the high-Risk Functions and Conditions.
2. Testing is performed throughout the entire SDLC and should validate Documents, Procedures and Code.
3. Changes to Requirements should be reviewed as thoroughly as the initial Requirements.
4. It is not practical to test sufficiently to eliminate all Defects.
5. Software testing involves effective Planning, Executing, Controlling and Closing.
6. It is better to identify Defects early in the process.
7. Each Test Case should involve a clear test transaction, documented entry procedures and observable Expected Results.
8. All forms of Conditions should be tested: business and technical, positive and negative, expected and unexpected results, and extreme boundaries.
9. Test Cases should be designed so that they can become part of a reusable Test Library.
10. Testing is a valuable skill that can be learned through education and experience.

II.   KEY PLAYERS IN THE TESTING PROCESS

As described in the previous chapter, all software should go through a Systems Development Life Cycle (SDLC) process, whether it is a Change to an existing Application or a completely new Application. The life cycle is essentially the same for both of these efforts.

Depending on the size and complexity of the Project, there may be multiple Development Teams, multiple Test Teams and multiple End User departments, but their activities are essentially the same in each case. For a small Project, one person may perform multiple activities.

However, each group involved in the Development process has an impact on the testing effort and its success. For the purpose of this discussion, we will assume that each of the activities required by the SDLC is covered by one of the following four groups of people:

- Project Management Team
- Development Team
- Test Team
- End Users

Each of these groups is discussed in the balance of this chapter.

The Project Management Team

The Project Management Team initiates the Project and oversees the effort through the entire life cycle. This Team, while not responsible for the day-to-day management of the Project, is responsible for the timely delivery of a quality Application that meets the End Users' needs. The Project Management Team should have representatives of each of the other teams: Development, Testing and End Users, and acts as an intermediary between the Project Team and interested parties, such as other Project Teams and senior management.

This organizational structure provides a forum that facilitates fluid communication between all of the appropriate parties regarding schedule changes, Project dependencies, and open Issues from a software and operations viewpoint. The Project Management Team is the core of any Project, and is the central Function that holds the Project together. The Project Management Team normally consists of several key players:

Project Manager

The Project Manager is responsible for all phases of the Application acquisition and/or Development and installation including Design, Development, Testing, Implementation, Conversion, Documentation and Training. The Project Manager ensures that the Project is delivered on time and within the Budget by coordinating the activities of the many teams that are necessary to make the Project successful.

Project Sponsor

The Project Sponsor is the senior manager who authorizes the Project. The sponsor approves Changes to the Application and significant Changes to the Project itself. The Project Sponsor may work as a member of the Project Team, but more typically receives updates and makes critical decisions.

Development Manager

The Development Manager is responsible for the Design and Development of the Application, performing certain Test Types, ensuring that a quality product is delivered to the Test Team, and supporting the Test Team through the testing process.

Test Manager

The Test Manager is responsible for Developing and Implementing the Test Plan, managing the day-to-day activities of the Test Team and reporting the results of the testing effort to the Project Management Team.

End User Representative(s)

It is critical to the success of any Project to have a representative(s) from the End User's organization(s) involved in all stages of the Project from Design, Testing and Implementation through to Training and Conversions. The End User Representative supports all phases of the development life cycle by providing answers to questions regarding Functions, suggesting Design possibilities, conducting certain Test Types, and acting as the liaison to the overall community of End Users.

Project Administrator

For Projects of all sizes, it is essential to have proper Documentation in all of the phases of the development life cycle. The Project Administrator keeps all the Documents that pertain to the Project, including:

- Functional and Business Requirements
- Meeting Minutes
- Project Plans
- Sign-offs
- Timesheets
- Expenditures, etc.

Systems Manager

The Systems Manager is responsible for running the computers and other hardware and software that support the Application while in Development, during Testing, and ultimately in Production. This includes ensuring that:

- Sufficient regions are available
- Sufficient PCs, printers, faxes and other personal equipment is available
- Back-ups and software installations occur as planned

Auditor

The Auditor monitors the Project to:

- Ensure that the Planning and Implementation conforms to all Regulatory Requirements and company Standards
- Verify that all security levels are observed for the Application and related processes
- Verify that contingencies are properly considered (e.g., Disaster Recovery)

The Development Team

The Development Team is primarily responsible for:

- Participating in the Requirements Phase
- Conducting the Design and Programming Phases
- Conducting certain Test Types

The Development Team has many roles, each of which is critical to the success of the Project, including:

Business Analyst(s)

The Business Analyst (BA) is responsible for the detailed description of the new Application or the definition of the Changes to an existing Application. The Business Analyst analyzes the current activity and reviews with the End Users how a new Application should work, or the potential Changes to an existing Application. The BA's business knowledge is used in writing the Business Requirements and Functional Requirements Documents. The Business Analyst also provides field mapping, data mapping and Workflow analysis for new products, Changes, vendor products and Conversions.

If the new Application is provided by a vendor rather than Developed in-house, the Business Analyst reviews all of the Functions in the vendor's product and determines where Changes to the product may be required to meet the organization's needs.

The Business Analyst provides support to the Developers and Testers throughout the SDLC by answering questions and making Changes as required through the Development and Testing Phases. The Business Analyst also maintains the Functional Requirements as a living Document so that it, at all times, reflects the End Users' needs.

Database Administrator

The Database Administrator (DBA) is responsible for designing the Database, including the fields and the structure that is needed for the Application or Change. The DBA uses the Design Document to determine what new fields and/or Tables may be required, and then works with the Business Analyst to Develop the Database to support the current Functions and potential future Functions to minimize the impact of future work on the Database.

The Database Administrator also works with the Developers to help them most effectively use the data and the structure of the Database to write the Code.

Developers

The Developers (Software Engineers, Programmers and Programmer/Analysts) are responsible for Coding the new Application or Changes to a current Application. The Developers may also work with other in-house technology groups, or with a vendor, when an Interface to another Application is needed to provide data either in or out of the Application that is being tested.

The Developers also conduct Unit Testing, Systems Integration Testing and Systems Technical Testing, and support the remaining testing through to Implementation. Developers make Changes throughout the Project in response to

Defects that were discovered during testing or Design Changes that are made during the Development of the Application.

Testers

Testers who are assigned to the Development Team are typically responsible for the Unit Testing and System Integration Testing, either on their own or teamed with a Developer. Depending on the size of the Project and the size of the Development Team, the Developers may perform this activity themselves.

The Development Team participates in all phases of the SDLC. The following summarizes the Functions the team performs in each phase:

SDLC ACTIVITY	FUNCTIONS OF THE DEVELOPMENT TEAM
Requirements	Participate in writing the Business and Functional Requirements for the Application or Changes
	Develop mapping for data and fields
Design	Participate in the selection of the Application or product
	Develop high level Design of the Application or Change
Programming	Code Application according to Design and Requirements
	Work with other Development teams to establish any required Interfaces to other Applications
Testing	Make Changes to resolve Defects or Design Changes
	Work with the Test Team to evaluate and prioritize Defects and support testing efforts
	Plan and Execute the Systems Technical Tests, including the Stress Test
Implementation	Participate in the roll-out of the Application

Figure 5 – Functions of the Development Team

The Test Team

The Test Team is primarily responsible for assuring that the software:

- Performs as intended
- Conforms to Standards
- Meets the End User's needs

The Test Team should participate in the Design Phase of the SDLC by reviewing the Requirements produced by the Business Analysts to ensure that the Application will conform to the corporate Guidelines and generally accepted Industry Standards for format, presentation, and navigation. The Test Team should participate in the Programming Phase by providing the Development Team with the information needed for Unit Testing and System Integration Testing.

The Test Team may be responsible for all or some of the following roles:

Test Section Managers

For a Large Project, the testing effort might have to be divided into multiple teams based on the Application's Functions or the company's organizational structure. The Test Section Manager's responsibility is to oversee the daily operations of their Test Team Section.

For instance, the Database Management Function may be split from the Accounting Functions for the operation of a Payroll Application. In this case it may make sense to have one Section Manager for the testing of the Database Management Function and one for the Accounting Function. Each Section Manager typically would have his/her own Detailed Test Plans that would be merged with the other teams' Plans to produce the overall Detailed Test Plan.

If the Project will make extensive use of automated testing tools, there should be one Section Manager assigned to manage the automation process.

Test Team Administrator

Just as with the Development Team, the Test Team needs to keep careful records to document all phases of the Test Planning and Test Execution. This includes the analysis that is performed to prepare the Test Plan, Test Procedures and Expected Results, as well as the Actual Results of the tests and the tracking of any Defects and their resolution.

All of this Documentation provides the necessary information for Management, Auditing and End Users to review the progress and quality of the testing effort. This Documentation, if properly maintained, will also make it easier to Plan tests for future Changes to the Application since much of the information, Test Procedures, etc., may be used again for Regression, Performance or Conversion Testing. The role of the Test Team Administrator is to manage all of the activities relating to this Documentation.

Authors' note: Save your work now, and save time later.

Testers

The Testers have many roles in the Test Preparation and Execution process.

- In the Planning Phase, the Testers write the Test Procedures, including the Expected Results.
- In the Execution Phase, the Testers Execute the Test Procedures (either manually or using an automated tool) and document the Actual Results.
- The Testers also record any Issues that arise from the testing, and work with the Development Team to resolve the Issues.

The Testers may also be called upon to review Training Documentation and User Manuals since they are often the first people on the Project Team to have hands-on knowledge of how the Application actually works.

Test Tools Implementer

The Test Tools Implementer has the responsibility for selecting and sometimes justifying the use of automated tools. With each new Project, automated test tools that were previously acquired should be re-examined to ensure that they can be used effectively within the current Test Project Plan and Budgetary constraints.

Once a tool is selected, the Test Tools Implementer manages the installation, use and maintenance of the test tool throughout the testing effort.

Tool Administrator

With any use of test automation, it is important to have a Test Team member who fully understands the day-to-day use of the automation tools. Depending on the number and types of tools, it may be necessary to have multiple Tool Administrators with different skill sets.

The Tools Administrator creates the library of automated Test Procedures that will be Executed during the testing process, determines what output is appropriate, such as performance metrics or calculations, and maintains all Documentation regarding the automated comparison of Expected Results to Actual Results. The Tool Administrator may work with the Development Team or the test tool vendors to make the tool work in the Test Environment.

The Test Team contributes throughout the life cycle by participating in the Design and Development Phases to ensure that all of the company's Testing Standards are adhered to, and in the Implementation Phase where they typically support Training, Conversions and the Implementation process.

The following summarizes the activities of the Test Team in the SDLC:

SDLC ACTIVITY	FUNCTIONS OF THE TEST TEAM
Requirements	Review Business Requirements for the Application or the Changes to ensure that they meet defined quality Standards
Design	Participate in the selection of the Application (if acquired externally) or platform
	Develop Master Test Plan to assist in establishing timelines and milestones
Programming	Provide support to Developers during Coding, and Defect resolution, by answering questions
Testing	Develop and Execute the Test Procedures as dictated by the Detailed Test Plans
	Identify and document Defects and work with Business Analysts and Developers to resolve and re-test them
Implementation	Participate in the Training, roll-out and Conversions that are required for the Implementation of the Application

Figure 6 – Functions of the Test Team

The End Users

The End Users play many essential roles in each phase of the SDLC. There may be representatives from multiple user areas, depending on how the Application is to be used. Each operational area that will be affected by the Implementation of the Application should minimally provide input to the End User Team, and ideally assign at least one experienced person to the Project.

User Experts

Some of the people assigned to the testing effort should be experts in the operational process that will be supported by the new Application or the Change. Their knowledge is invaluable in building Test Cases and in evaluating the Actual Results.

Testers

The End Users might also provide some people to act as Testers. These people might not know the business in detail, but they will be involved in using the

Application when it is in Production. By working on the Test Team they will assist with the testing and learn how to use the Application.

The following summarizes the activities of the End Users during the development life cycle:

SDLC Activity	Functions of the End Users
Requirements	Document the business needs that the Application should cover
Design	Assist with the interpretation of the Business Requirements as the Application Design is established
Design	Sign-off on the final Design for the Application
Programming	Provide support for the Development Team by answering questions on Business Functions
Testing	Provide the Test Team with realistic Business Scenarios for testing
Testing	Prioritize the Issues to be addressed
Testing	Review and sign-off on Test Results
Implementation	Develop Workflows for the post-Implementation organization
Implementation	Participate in the development of Training materials and Application Documentation

Figure 7 – Functions of the End Users

In the past, Developers were solely responsible for testing their own Code before it went into Production; however, it quickly became clear that a Developer is not the best person to test his/her own work. As Applications became more complex and more Developers were added to the mix, it became impossible to have each Developer test their own Code and still have an Application that works as a single unit. With this realization, the role of the independent Tester evolved.

The Testing Role

Understanding the testing role for Medium to Large Size Projects[2] has matured over the last few years to the point that almost everyone agrees that a dedicated Test Team is critical to the quality and success of the Application. The Test Team may be as small as one person or as many as 100, but the key point is that resources are dedicated to the testing effort.

> **Authors' note:** Medium and Large Projects must have dedicated testing resources.

---

[2] A process to estimate the size of a Project is presented in Chapter V.

Each company may place the Test Team in different organizational structures, generally either as part of the End Users' organization or as part of the Development Team. Each of these approaches has advantages and disadvantages.

	ADVANTAGES	DISADVANTAGES
PART OF END USERS' ORGANIZATION	Ensures that the Application meets the business needs.	End User organizations may lack the expertise and/or Training in testing methodologies and tools
		Qualified Testers may be among the most experienced people in the End User organization, and may be needed on multiple Tasks simultaneously
		Assigning testing to the End Users can create unnecessary conflicts when Defects arise from poorly defined Requirements
PART OF THE DEVELOPMENT ORGANIZATION	Allows Control of quality to be in a single organization	Management of the testing effort is often a full-time job
		Testers may not be comfortable identifying Code-related Defects within their own organization
		Management of these two teams can result in divided loyalty issues

Figure 8 – Pro and Con of Testing Team Reporting Structure

The structure that seems to have worked best for Medium and Large Projects in Financial Services Firms is to have the Test Team report directly to the overall Project Manager, regardless of whether that Project Manager is part of the End User organization or the Development organization.

This allows for a relatively independent Test Team that will ensure consistent quality Standards across all segments of the Application, regardless of how many organizations are involved. The Project Manager is ultimately responsible for the success of the Project, and must have the autonomy to identify and report on Defects, regardless of where they occur in the organization.

Team Support Roles

The roles outlined above are important to the success of the overall Project, and they are critical to the success of the testing effort. The testing effort cannot be successful without the active support of each of the Management, Development and End User Teams.

The Test Team relies on the Project Management Team to provide the following:
- Definition of the Scope of the Project
- Timeline and milestones for the Project
- Budget/resources to support testing effort
- Communication of the Project status to the End User Team and other managers

The Test Team relies on the Development Team to provide the following:
- Business and Functional Requirements that outline in sufficient detail the features and Functions of the Application
- Quality Code that is ready for testing, and that conforms to the Business and Functional Requirements Documents
- Resolution of all open Defects/Issues through timely turnaround of quality Code
- Assistance with automated tools for testing

- Assistance with running of the Application software until it is moved into Production

The Test Team relies on the End Users to provide the following:

- Clear definition of the business needs that are to be met with the Application
- Assistance in formulating realistic Business Scenarios
- Assistance in prioritizing the Business Scenarios
- Conduct User Acceptance Testing and Parallel Testing to ensure that the complete test cycle is Executed
- Assistance in prioritizing the resolution of Defects

Although each of the activities outlined above is important in some situations, it may be that not all of them are necessary or practical for a specific Project. For Small or Medium Projects, some of the activities can be combined or eliminated.

Summary

This chapter identified the following key points:

- There are several teams involved in testing and different roles within the teams, including:
    - Project Management Team
        - Project Manager
        - Project Sponsor
        - Development Manager
        - Test Manager
        - End User Representative(s)
        - Project Administrator
        - System Manager
        - Auditor
    - Development Team
        - Business Analyst(s)
        - Database Administrator
        - Developers
        - Testers
    - Testing Team
        - Section Manager(s)
        - Test Team Administrator
        - Testers
        - Tools Implementer
        - Tools Administrator
        - End User(s) assigned to the Testing Team
    - End Users
        - User Expert(s)

- Testers
- Each team has an impact on the other teams and is responsible for certain aspects of the Project
- Medium and Large Projects require a dedicated Test Team
- The decision regarding where the testing effort should reside in the organization can influence the outcome of the Project

III. TEST TYPES

Every new Application or Change to an existing Application requires some level of testing. Although testing could be conducted by one group, usually several teams participate and own different aspects of the testing effort. In the examples in this chapter we assume that three separate teams perform the described tests:
- The Development Team
- The Test Team
- The End Users

This three-team structure most often exists for a Large Project with multiple Applications and/or a high level of complexity. By using the diverse expertise of each team, this structure offers the best opportunity for comprehensive testing.

There are many different Test Types that are available, and which can be used in different circumstances. This chapter examines the various Test Types that are most commonly used in testing software for the Financial Services Industry by reviewing five groups of tests.
- Requirements Reviews
- Tests performed by the Development Team
- Tests performed by the Test Team
- Tests performed by the End Users
- Special Testing Situations

These five groups are discussed in the following sections.

Requirements Reviews

Before any of the Coding begins, the Requirements and the Design should be reviewed. These Structured Reviews, also called Inspections or Walkthroughs, are actually the first tests that should be performed, and can take place in three forms:

Business Review

The Business Review of the Requirements takes place by gathering all of the interested parties from both a technical and business perspective and discussing each aspect of the Application's Functions as described in the Business and Functional Requirements Documents. Each of the participants will have their own view of the Application, and the review should answer the following questions:
- Is this a reasonable solution for the business need?
- Does the process defined by the Requirements meet the business need?
- Are all of the Functions available?
- Are the Functions consistent with corporate Standards regarding presentation, navigation and reject repair?

When each of these questions has been sufficiently answered and all of the parties agree on the Functions, the Application can be Designed.

Technology Design Review

After the Business Review has been signed-off and the technical Application Design has been completed, the Design can be reviewed by knowledgeable people, independent of the Designers. This review should include questions such as:

- Does the Design meet the company's technical Guidelines and Standards?
- Will the Design work within the company's existing technical architecture?
- Does the Design take advantage of the available technologies?
- Is the Design possible from a technical standpoint?
- Does the Design support the Projected volumes?
- Does the Design allow for future Changes and Maintenance?

When the review of the Technical Design has been completed, the Data Design can be performed.

Data Design Review

After the Data Design has been completed, a group of the appropriate technical experts should perform a Walkthrough of the Requirements Documents and the Design Document to ensure it will support the required Functions. The following questions should be answered as part of this review:

- Does the data support all of the Functions in the Business Requirements and Functional Requirements?
- Are all of the data relationships correctly defined?
- Is the naming convention consistent with corporate Standards?
- Does the Data Design work with the corporation's other Applications?
- Is the retention and purge process defined and acceptable from the business' perspective?
- Does the restart and recovery process meet corporate Standards?

These reviews of the Business and Functional Requirements and the Application Design are important steps in the Systems Development Life Cycle. Because these activities establish the foundation for everything that follows, it is important for the Test Team to participate throughout the Requirements and Design and Phases.

This process allows Functional gaps, logical inconsistencies and Design Issues to be resolved before the Coding is started, and it helps the Test Team understand how the Application should work as well as address any End Users' concerns.

Testing Performed by the Development Team

The Development Team has responsibility for defining and delivering high quality Code that meets the End Users' needs. To do this, there are three main Test Types that are performed on the Code prior to that delivery:

- Unit Tests
- Systems Integration Test
- Systems Technical Tests

Each of these Test Types is discussed in the following sections.

Unit Tests

Developers normally perform some informal testing, called Desk Checking, on their Code throughout the Development process. When they feel that a unit of

Code is correct, they perform a more formal Unit Test, which is intended to ensure that the unit is:

- Functionally Complete
  - Includes all of the defined Functions
  - Exception Conditions and Error Handling are available
  - Error Recovery Procedures are available
- Defect Free
  - Statements, Conditions, logic and paths perform correctly
  - Internal logic performs correctly
  - Internal Design supports the Requirements and business purpose

Unit Testing is the process of formally testing the smallest operating unit of Code. For instance, this may be one logical Module, one Screen or one Report, and is not dependent on input or output from other units of Code. The objective of this Test Type is to ensure that each unit of Code works properly when exercised independently.

Unit Testing is also called White Box Testing. In White Box Testing, the Developer can "see inside the box"; they can provide input and follow it step by step through the unit of Code that is being tested. In the Test Types performed by Testers and End Users, called Black Box Testing, the Tester does not have access to the Code, and they cannot "see inside the box". White Box Testing and Black Box Testing are described more in Chapter XIII – Managing Testing Automation.

Unit Testing is most often performed by the Developer, who artificially inputs the required data into the unit of Code and steps through the Code to ensure that the logic correctly responds to the input. For example, if the unit of Code being tested updates a Database, the Developer inputs data directly into the program and watches as the data is added to the Database. The Code verification is performed by separately querying the Database and by validating that all of the updates are correct.

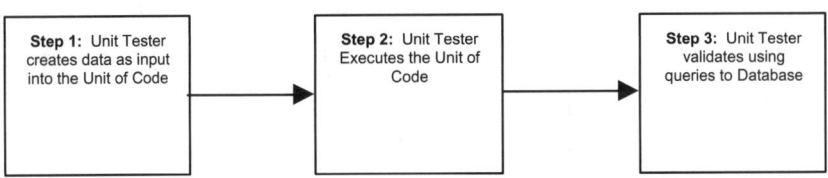

Figure 9 – Functions of Unit Testing

System Integration Test

System Integration Testing combines the smaller units of Code that were tested in Unit Testing, and exercises the combined Functions logically while continuing to use the White Box approach to testing. The objective of this level of testing is to ensure that all of the smaller units of Code work together as a cohesive Program, and that the flow of data through the Application is consistent with the Design that was defined in the Business Requirements or Functional Specifications.

Developers usually perform a sub-type of the integration test, called an Incremental Integration Test, where units of Code are added after previous units of Code have successfully worked together. This incremental approach allows Developers to isolate Code Defects as they exercise their work.

The System Integration Test allows the data to be entered as it would be entered into the Application in Production, rather than by simulating the data input through Database manipulation. This allows the Tester to validate the operator procedures, Warning/Error Messages, panels, Reports and recovery procedures.

In the case where data is coming from another Application, either through Messaging or Interfaces, it may be necessary for System Integration Testing to simulate that data, depending on the availability of data from the sending Application. This allows the Tester to validate relevant programs independently of outside Applications. When data from the sending applications becomes available and real circumstances can be reproduced, the Tester can validate the Interfaces, Message content and format, and recovery procedures.

The System Integration Tests should mirror the Conditions and Environment that will be used by the Test Team when they begin to test. However, for both Unit Testing and System Integration Testing, the Testing Scope is limited. These Test Types do not typically cover the universe of Conditions, positive and negative, that could occur within the Application. More exhaustive testing is normally covered either in Business Functional Testing or Business Technical Testing.

Overall, the Systems Integration Test is intended to validate that the Application:

- Performs correctly and reliably within the appropriate time constraints in the intended environment
- Interfaces correctly with any other appropriate Applications
- Is technically sound and has functioning recovery capabilities

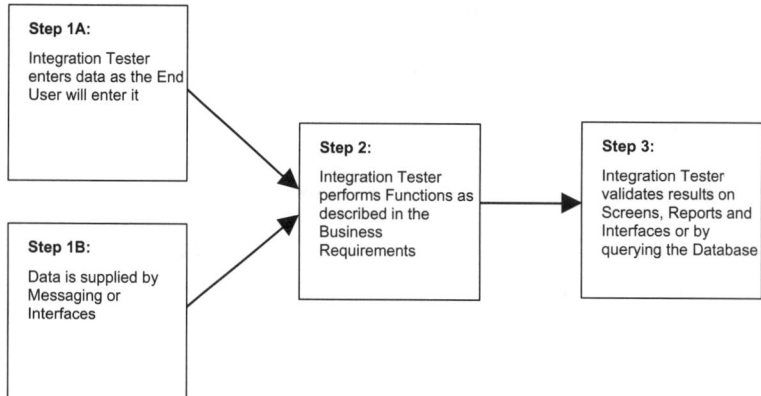

Figure 10 – Functions of System Integration Testing

The Unit and System Integration Tests are typically completed before the Code is delivered to the Test Team. These tests are intended to ensure that a working, high-quality Application is delivered to the Test Team so that they may begin testing without the complications of units of Code that do not work together.

Another activity that occasionally is performed as a part of the Systems Integration Test is a Walkthrough of the Code. In a Walkthrough, a group of experienced Developers reviews the actual Code by examining it and talking to the Programmer about how it is intended to support the Functional Requirements. This process can identify areas where the Code does not do what it is supposed to do, or is inefficient.

Systems Technical Tests

In addition, there is another type of testing that is normally performed by the Development Team that does not necessarily need to be completed prior to the delivery of the Code to the Test Team. Called Systems Technical Testing, some of this testing may be performed concurrently with the Test Team's testing or may be done prior to the delivery.

Systems Technical Testing is managed by the Development Team, with or without support from the Test Team, and can include several different activities:[3]

- Code Coverage Test
- Compatibility Test
- Compliance Test
- Conformance Test
- Contingency Test
- Disaster Recovery Test
- Interoperability Test
- Recovery Test
- Security Test
- Stress Test (Including Load, Volume, Performance and Storage Tests)

Stress Testing is one of the most visible types of Systems Technical Testing and it validates that the Application will successfully process high volumes in a timeframe that will be satisfactory to the End Users, and will fit within the constraints of the timetables of normal and peak business volumes. There are several types of volume Conditions that may be covered in Stress Testing:

- High volumes of users performing concurrent activities
- High volumes of Messages in and out of the Applications
- High volumes of on-line transactions
    - In a specific period of time
    - Over a sustained period of time
    - Just prior to a cut-off time

> **Authors' note:** The Project Manager must ensure that an Application can handle anticipated Production volumes.

- High volumes in batch processing within a narrow processing window
- Large data bases with complex queries in narrow timeframes

The volume level of the transactions that are tested should be some multiple of the highest level that is expected during the life of the Application. The peaks should reflect the Application's ability to handle large volumes over a sustained period of time, as well as its ability to handle peaks that occur just before cut-off times. This should include large numbers of simple transactions as well as large

---

[3] These terms are defined in the Glossary.

numbers of complex transactions that exercise multiple processing Modules and require numerous data look-ups.

Validating that the Application will perform adequately is critical prior to a period of high volumes, but may not necessarily prevent the initial Implementation of a new Application or a Change. For instance, if a new Client Reporting Application is to be installed in February and the high volume period is at the calendar year-end, then a decision may be made to allow the Application to be Implemented without performing the Stress Test because there is enough time remaining to allow for a Stress Test and performance upgrades if necessary.

The Stress Test may be Executed in many ways. Often, an automated test tool is used to simulate high numbers of simultaneous users or to create high transaction or Message volumes. The automation of Stress Testing often allows a more accurate record of the response and run times. In most cases, the type of data used for Stress Testing is less of a concern, since the key objective of the test is to measure the performance under high volume Conditions.

Testing Performed by the Test Team

Once the Code is delivered to the Test Team, the Application may go through multiple Test Types. Again, the Test Types that are selected for the Application will depend on the size and complexity of the Application, and the organizational structure of the Project Team. The Test Team is responsible for certifying that an Application performs according to the defined Functions, and that it is ready to be delivered to the End User for testing or Implementation.

To do this, the following Test Types may be used:

- Business Technical Tests
- Business Functional Tests
- Business Integration Test
- Regression Tests

These four Test Types are described in the following sections.

Business Technical Tests

The purpose of the Business Technical Tests is to exercise the Code from several viewpoints and answer the following questions:

- Is the Code that was delivered the Code that was expected?
- Is the Code stable?
- Does the Code perform correctly under various data Conditions?
- Does the Code Interface properly to other Programs and Modules?

The Conditions that are normally exercised in this type of testing do not focus on business Conditions or business logic, but rather on the positive and negative data Conditions that are processed in the Application's connections. See Chapter X for more information on the common testing Conditions that are related to Business Technical Testing.

Business Functional Tests

Functional Testing examines smaller operational units of Applications that can be tested independently, before they are combined for a complete end-to-end

Business Integration Test. Functional Testing may be designed so that some testing can begin as soon as the first operational unit of Code is available, rather than waiting until the entire Application Development is complete.

Functional Testing exercises the operation of the Application by relying on the User Interfaces, and validates the Application by using the available stable Reports, Screens and Application Interfaces.

Functional Testing differs from Unit Testing in two ways:

- Unit Testing focuses on a unit of Code or a small group of Modules. Functional Testing focuses on an operational unit, which may be comprised of multiple Programs or multiple groups of Modules.
- Unit Testing is White Box Testing, while Functional Testing is Black Box Testing. The Tester cannot see how the Code works, so the approach is from a business or operational standpoint. The Tester can only say that the Actual Result is the same as or different from the Expected Result, and the Developer must use this information to locate the reason for the Defect in the Code.

Functional Testing differs from Business Technical Testing based upon the type of data that is used. In Functional Testing[4] the Test Team uses Test Transactions, based upon the Business Scenarios, which reflect how the Application will be used by the business. The process of developing tests is discussed in Chapter IX.

One of the major benefits of performing a Functional Test is that operational units of Code can be exercised independently so that Defects can be isolated to a limited area of the Program. Once one operational unit is stable, the next unit of Code can be added to the test to build on the first unit, thereby increasing the opportunity to clearly identify the sources of Defects.

Business Integration Test

Business Integration Testing, often called End-to-End Testing, brings together all of the Application's Components and performs testing that covers all of the Application's Functions from initial data entry through to posting in all of the dependant Applications and through to Reporting.

> **Authors' note:** Business Integration Testing should ensure that all of an Application's Functions work together as required.

The Business Integration Test is the first time the entire Application is tested in a Production-like mode and all of the parts of the Application are available for testing. The Functional Test should have been performed prior to this, and the Code should be reasonably stable for this test, allowing the focus of the tests to be on the Business Scenarios.

The objective of Business Integration Testing is to ensure that all of the of the Application's Functions work according to the Requirements, from the initial inputs to the final Reports, and that it will handle not only the normal Business Scenarios, but also those that are less likely to occur. During this test there should be no data that is entered other than how it would be entered in Production.

---

[4] The same Test Transactions should also used by the Business Integration Test and the User Acceptance Test, which are described later in this chapter.

In the Business Integration Test, all of the Changes to outside Applications that support the Application should also be in place so that the test can accurately reflect the Environment, the Interfaces and the Code that will be in place when the Application is Implemented.

Often, the Business Integration Test supplements the Manufactured Data with real Production Data to mirror specific Conditions. Validation during this type of testing focuses primarily on the results the End User would see such as Reports, Screens and Interfaces to other Applications; however, it may also be necessary to use queries to validate some categories of data.

Regression Tests

Regression Testing validates that any Changes that were made to other parts of the Application have not affected the rest of the Code. Regression Testing can be included in the testing of a new Application, or after the Application has moved into Production and is modified. There are many types of Projects where a Regression Test may be necessary:

- Phased testing of an Application, where previously tested Functions need to be retested as new Functions are added
- Enhancements to an existing Application
- New technical platform for an existing Application
- Vendor Releases for Applications that were purchased from a Application vendor
- Changes to an Application that sends data to the new Application
- New core data such as Code Sets or Cross Reference Tables
- New sources of data

In each of these situations, any new Functions could be tested in Unit, Systems, Functional and Business Integration Testing, but the Regression Test will ensure that there is nothing in the new Functions that will negatively impact the previously tested Application.

A sub-set of Regression Testing is called Changes Only Testing. In Changes Only Testing, as the name indicates; only the Changes themselves are tested. In Regression Testing, the Changes are tested as well as any related activities.

The Regression Test is usually accomplished by re-running the Test Cases and the Test Procedures that were developed during prior testing efforts or other Testing Phases. This type of testing makes the best use of a Test Library and automated testing tools, as will be covered in Chapter XIII.

Testing Performed by the End Users

The End Users may be responsible for managing the following Test Types, or the Test Team might manage them. In either case, the End Users must be very active in supporting the tests in order to obtain the most benefit.

- User Acceptance Test
- Parallel Test
- Conversion Test

These three Test Types are discussed in the following sections.

User Acceptance Test

In order for an Implementation to be considered successful, the End Users must be satisfied that the Application has been fully tested and is performing as expected. To obtain this level of satisfaction, the End Users should participate in the Test Planning, review the Test Results and/or participate in the Testing Phases performed by the Test Team, and conduct their own independent testing. When the End Users perform their own testing, this test is most commonly referred to as a User Acceptance Test.

The User Acceptance Test is usually designed by and Executed by the End Users. Very often, the testing is a mix of the normal Business Scenarios that are performed by the operational area and the more unusual situations that the End Users have encountered in Production. User Acceptance Testing can be performed at two levels:

- Perform an independent Business Integration Test, using either the Test Team's Test Transactions or their own
- Use the Application entirely as it would be in Production without the aid of test-related queries or testing automation, often by entering real Production Data

Either way, a UAT should consider the following points when it is designed, although not all of these Conditions must be included in every UAT:

- Are all of the Application's Functions covered by the tests?
- Is every organizational unit that will use the Application represented?
- Does the End-of-Day run properly?
- Does the Application perform properly over multiple days?
- Does the Application perform properly over a weekend if there are separate weekend Functions?
- Does the Application perform properly at month-end, quarter-end or year-end?
- Can a day's work be performed in a day?
- Can the Application support peak processing just prior to a cut-off time?
- Can the Application support error processing (reject reentry) in the time available?

The User Acceptance Testing should also be designed to serve three other purposes.

> "I hear and I forget.
> I see and I remember.
> I do and I understand."
> Confucius

Training

A properly designed UAT allows users to learn the Application in a Test Environment without having an impact on Production. The pace of the UAT is usually slower than Production, so the users can become comfortable with their knowledge of the Application without the pressures of daily processing.

Workflow Validation

The UAT allows the End Users to test modifications to the Workflow that result from the new Application or Enhancements, again without the pressures of the Production Environment. By testing the Workflows prior to Production, changes can be made where they are needed without impacting Clients. This validation process is often called by many different names:

- Model Office Testing
- Operational Testing
- Procedures Testing
- Process Flow Testing
- Simulated Work Environment
- Usability Testing
- White Room Testing

User Documentation Review

The End Users should review the User Guides and any User Training material during the UAT to ensure that it is correct and that it is useable by people who are not involved in the actual testing.

The design of the UAT will have to balance the following factors:

- The desire to fully test every Condition prior to going live
- The pressure to go live as soon as possible
- The desire to have every user gain some processing experience on the test Application prior to the Application moving into Production
- The impact on current Production of having multiple users actively involved in the UAT

Parallel Test

For Projects that are replacing highly complex Applications with significant operational or financial Risk, the End User may feel it is necessary to run both the new Application and the old one simultaneously for some period of time to ensure that the Actual Results are the same from both Applications. Responsibility for this test is typically shared between the Test Team and the End Users, although the End Users usually have the primary responsibility.

> **Authors' note:** A Parallel Test should be performed whenever there is a potential operational or financial Risk with a new Application.

In a Parallel Test, the same Production Data is entered into both Applications and all of the processes are run in both, including batch processes, Client Statements, and on-line queries. The objective of the Parallel Test is to validate that the results from both the current and the future Applications will be the same, or verify that any expected differences did occur. As with the User Acceptance Test, the Parallel Test allows for a validation of the Workflows in a more Production-like Environment and helps to Train the users for the new Implementation.

It is not uncommon during a thorough Parallel Test to discover Defects in the current Application. In this case, the users will have to decide whether to notify their Clients of changes to the processing rules that are being used, or to modify the new Application to mimic the current Application.

Conversion Test

Typically with a new Application that is replacing an existing one, and occasionally with significant Changes to existing Applications, data must be converted from the old Database (or Files) to a new Database. The data that can be converted includes static data such as Tables, Master Files, etc., and Dynamic Files such as positions, transactions, etc. The Conversion and reentry of this data can be performed manually, through automation, or a combination of the two. Responsibility for conversion testing is also often shared between the Test Team and the End Users.

There are two categories of tests that are required for Conversions:

Conversion Process Test

If the process involves any form of automation, the Program that is used to convert and move the data should follow a normal SDLC process and should undergo its own separate testing.

Converted Data Test

Whether data has been moved manually or through automation, the data in the new Database must be validated after the Conversion, and before Production begins, to ensure that is it correct.

Special Testing Situations

In addition to the basic Test Types described in this chapter, there are two special types:

Comparison Testing

If a firm is considering acquiring an Application from a vendor, there is usually a need to compare more than one candidate. In our experience, this type of test is performed by Developers, End Users or a consultant, and is best performed with the following steps:

- Identify alternative vendors
- Prepare a detailed list and a summary of the Functions and technical considerations
- Categorize these Functions and considerations as mandatory and optional
- Send the summary list to the appropriate vendors
- Contact existing users of the Applications
- Evaluate the responses and prepare a short list of three to five vendors
- Add weights to the various questions [5]
- Send the detailed list to each short-listed vendor
- Calculate the weighted total of each vendor's response
- Invite two to three vendors to make detailed presentations
- Select one vendor and perform a mini-Functional Test at the vendor's location

> "Real knowledge is to know the extent of one's ignorance." — Confucius

---

[5] The authors' weighting process is described in Chapter V – Getting Started.

- Select the tested vendor, or test another and compare the Test Results

Most of the steps in this process involve judgments based upon experience. Since the effort required to actually run tests in a vendor's Application will exceed the capacity of most firms and most vendors, it is necessary to perform a manual comparison of the vendors' claims before hands-on testing.

Prototype Testing and Pilot Testing

Prototype Testing is usually performed on an Application that is being Developed either internally or externally. The concept behind a Prototype Test is to see if the Function and presentation of the Application seems to be what the End User's want. If the Prototype Test is considered successful, the Development of the Application will probably continue.

A Pilot Test differs from a Prototype in that a Prototype has very limited Function.

- A Pilot usually has a few Screens that perform all of the related Functions.
- A Prototype usually has many Screens that show the breadth of the Application, but which does not perform the Functions, it merely simulates them.

Summary

Every Project is different and will require an assessment of the Test Types that are needed so that everyone involved feels that the testing is thorough and accurate, and that the Application is ready for Implementation at the end of the testing. By making it clear at the beginning of the testing effort that the people responsible for testing the Application will have to personally sign-off on the new Application before it goes live, the process should be more thorough.

Determining which of the Test Types will be used for the Project will depend on many factors, not the least of which are the availability of resources and the time permitted for testing. Chapter IV will explore the differences between each type of Project and then Chapter V will identify how testing can be started. Chapters VI, VII and VIII will walk through the process of building the Test Plan. It is during the development of the Detailed Test Plan that the Test Types that will be used are identified and expanded so that each of the participants in the testing process can begin to prepare for their own testing responsibilities.

This chapter also identified the following key points:

- There are a number of different Test Types that can be performed in various situations.
- The Requirement Documents should be reviewed by groups of experts before Programming begins. These reviews should include:
  - Business Review
  - Technical Design Review
  - Data Design Review
- There are a variety of different tests that must be performed by the Development Team to ensure that they deliver a Functioning Application:
  - Unit Tests

- Systems Integration Test
- Systems Technical Tests
* There are several Test Types that must be performed by the Test Team to validate certain kinds of Applications:
    - Business Technical Test
    - Business Functional Test
    - Business Integration Test
    - Regression Test
* There are three Test Types that must be managed by the End Users to validate certain kinds of Applications, although they can be performed by the End Users, the Testers, or a combination of both:
    - User Acceptance Test
    - Parallel Test
    - Conversion Test
* The UAT also helps the users learn how to use the Application and to exercise any new Workflows that will be needed.
* Some special types of tests exist that can be used in certain situations. These tests are:
    - Comparison Testing
    - Pilot Testing

IV. TEST PROJECT CATEGORIES

During the Design Phase, as the Master Test Plan is being initiated, it is important to identify the Test Project Category. Once the Test Project Category has been established, the Test Manager can define the Scope of the test effort based upon the size of the Project[6], decide which of the Test Types[7] will be required, and continue to build the Master Test Plan[8].

The Test Manager can identify the Test Project Category by asking the following questions:

- What is to be tested?
  - Is it a New Application?
  - Is it a Change to the Code?
  - Will it involve a Conversion of Data?
- Who is writing the Code for the Application/Change?
  - Will it be Developed In-House?
  - Will it be Purchased from a Vendor?

Based upon the answers to these questions, there are five possible Test Project Categories:

	NEW APPLICATION	CHANGE	CONVERSION
BUY FROM A VENDOR	Category 1	Category 3	Category 5
BUILD IN-HOUSE	Category 2	Category 4	

Figure 11 – Test Project Categories

The Test Types that will be required will differ based upon whether the Project is 'build' or 'buy', and whether the Project involves Changes to an existing Application or building an entirely new Application. Where there is a new business need, there are a number of Components that should be considered in order to ascertain the Test Types required. In all cases, the first step is to understand what is needed by the business.

Understanding the Business

The Test Team must understand:

- Each Function as it is described in the Business and Functional Requirements
- The business purpose and how the Functions will be used, including:
  - Use of the Application
  - Related Workflows and operational procedures
  - Client or Regulatory expectations.

The testing effort will need to cover all of these activities in at least one phase of the testing.

---

[6] A process for sizing the Project is described in Chapter V.
[7] The various Test Types were defined in Chapter III.
[8] The Master Test Plan is discussed in Chapter VI.

Understanding Interfaces with Existing Businesses

In most cases, the new business will not operate independently in the organization. Any new activity will most likely Interface with at least the General Ledger and/or Reconciliation areas of the organization. These Interfaces may be manual or automated. In either case, the Test Team and End Users must understand the interfacing units' expectations to ensure that the Testing Scope covers their concerns as well as those of the core business unit.

In the rest of this Chapter, we will discuss the Test Types that are required to support each of the six Test Project Categories.

Testing New Applications (Categories 1 and 2)

A new Application could be an entirely new Application or it could replace an existing one. There are many business situations that could lead an organization to decide that a new Application is needed. Some examples are:

- The organization is entering into a new type of business; for instance, if a brokerage firm decides to begin securities lending, it may decide to obtain a new Application, rather than update the existing Applications to process this business because it is handled differently from its other securities transactions.
- The organization decides to move from a legacy front-end to web-based Applications for easier delivery of data to users.
- The existing Applications are too expensive to run from an operations or technical standpoint, and an opportunity exists to move to less expensive technology.

Once the decision has been made to Implement a new Application, the Project Management Team should consider whether the Application should be Developed in-house or if an existing Application should be purchased from a vendor. This decision is made based on a number of factors:

- Are there any Applications on the market that meet the need?
- Are there skilled people in the organization that could develop the Application?
- What is the cost comparison between building and buying?
- What is the timeframe for Implementation?
- Is there an existing corporate pre-disposition to build vs. buy?
- What is the potential to integrate a vendor Application into the existing architecture?
- What is the potential to use the new Application as a foundation for an upgraded architecture?

Category 1 - Testing a New Application from a Vendor

There are many reasons why an organization may choose to buy or lease a vendor Application as opposed to building the Application using in-house Design or Development resources, as described previously in this chapter. If an appropriate vendor Application exists, the two most common reasons for selecting it are time and cost.

- A vendor Application often can be Implemented faster than in-house Development, so if time is a critical factor in the decision and an Application is available to meet the organization's needs, a vendor-supplied Application may be the right choice.
- The cost of a vendor Application can also be significantly less than in-house Development, with the trade-off that most vendor Applications do not meet an organization's exact needs and must be modified to some degree, or compromises must be made.

Alternatively, if a suitable Application is not available, or if the available Applications require extensive Changes, the decision might be to build, either in-house or using an external Development organization. This is covered as Category 2 later in this chapter.

Each of the following factors must be explored in order to determine the level of testing that is needed for an Application acquired from a vendor.

Source Documentation

To start Planning the testing of the Application, the vendor should provide Documentation so that the Testing Scope can be fully understood. There are varying levels of Documentation that can be supplied under the vendor contract, but in a best-case situation, all of the following pieces will be available:

User Manuals

The User Guides or Training Materials should provide samples and editing criteria for all of the workstation's Functions, detailed set-up requirements for all Functions and samples of all Reports. This will provide valuable information in Test Planning, and in Training the testing staff on the vendor Application. This information may have been initially prepared by the vendor, or by the End Users.

Business or Functional Requirements for Custom Code

Any Function that is being programmed specifically for an organization should have Business Requirements and Functional Requirements available as a source for Test Planning. These Documents may be written internally, by the vendor, or by a combination of staff. Before any Coding starts, the Requirements Documents should be approved by the Project Management Team and by the vendor (if appropriate).

Code-Values or Rule-Sets

For Functions that are driven by hard-coded and/or user defined Code-Values or Rule-Sets, a list of all the codes and/or rules, their allowable values and their processing implications should be provided to the Test Team. The Test Team will also need the same Documentation for the current Application, and ideally a map between the two Applications will be available. If not, it will have to be created. These Documents will be used to better understand the Testing Scope of the Code-Values and Rule-Sets.

> **Authors' note:** The vendor must provide a contact person(s) who knows the Application and who is available to help.

In addition to these Documents, it is also important to have pre-defined vendor contacts who understand the Application in depth and who are available to assist in the Planning of the testing effort and in the evaluation of the Defects that are discovered. For large and complex Applications, or Applications that are mission critical, the Project Manager should insist upon on-site vendor support.

Some other categories of Documentation that could be useful are:
- Programmer Specifications
- File Layouts
- Data Flow Documentation
- Table Structure of the Database
- Workflows and Operating Procedures
- Vendor Documentation Release Notes
- Marketing Documentation that describes the Application
- Test Cases that are already in the Test Library
- Testing Plans and Results from Previous Releases

Analyzing Vendor Applications

In addition to reviewing any Documentation supplied by the vendor, the Test Team must consider other sources of information in order to analyze the Application.

Evaluating Similar Clients

No two organizations are the same, and no two users of a vendor Application are the same. However, it is important to understand the types of organizations that are already using the vendor Application that has been selected. For instance, if a selected Custody Application is currently only being used by Clients with domestic securities operations and the acquiring business has a global component, the majority of the testing may focus on scenarios related to the global side of the business.

The best source of information about existing Clients is the vendor's user group if one exists. If it doesn't exist, the Project Manager will have to obtain a list of their current Clients and try to determine how those Clients are using the Application. There are a number of areas that should be compared in order to determine the similarity and/or differences with the Client base using the Application:

- Size
  - How does the Client base compare to the size of the organization in transaction volumes, user volumes, etc.?
- Type of Business
  - Are there other Clients that have a similar core business; i.e., Global vs. Domestic, Whole Life vs. Term insurance, Daily vs. Monthly Accounting, etc.?
- Organizational Structure

- If the use of the Application mandates or assumes some organizational structure, how does this differ from the firm's existing structure?

If the acquiring organization is the first of its size, type or structure to use the Application for certain activities, it will be important to focus the testing on the areas of difference and make some level of assumptions regarding the reliability of the Application where the rest of the Client base is similar.

The first Client, or the first major Client, will have to conduct an extremely thorough set of tests before moving the Application into Production.

Evaluating Application Complexity

Complex Applications require a detailed understanding of the Business Requirements and of the Application's internal processing logic in order to develop comprehensive test scenarios. The more complex the Application, the more detailed the tests.

Evaluating Interface Requirements

Some Applications require a large number of Interfaces to other Applications. In this case, each Application depends on the data that is processed through the other Applications, and the Test Cases must consider all of the potential combinations of data, including the processing of cancels, corrections and rejects.

Vendor's Internal Testing

Every vendor will test its Application prior to delivery, but the level of testing varies greatly across vendors. For this reason, it is important to understand how the Application is tested before the vendor releases it to its Clients.

Depending on the relationship with the vendor, there may be different levels of information they will supply regarding their testing of the core Application, Maintenance releases and custom Code. Some vendors are very willing to share this information with their Clients, while others refuse to disclose the results of their tests and their list of known Defects.

> **Authors' note:** If a vendor refuses to share their testing information with you, you should suspect their level of testing.

In order to understand the testing the vendor has performed internally, there are three key areas that should be reviewed:

- Test Organization
  - Does the vendor have a dedicated Test Team and if so, how is this area organized?
  - To whom does the Test Team report within the vendor organization?
  - Does this person have the authority to stop a release if it does not meet their Quality Standards?
- Test Methodology

- Are there multiple Test Types performed (i.e., Unit, Functional, etc.)?
- Are there multiple test cycles (i.e., Daily, monthly, contract life, settlement periods. etc.)
- What is being tested (i.e., stand-alone Application, integrated, etc.)?
- What is the source of their test data (Production Data, Manufactured Data, or both)?
- Do their existing Clients participate in any testing conducted by the vendor?

• Test Documentation
- Are the Test Procedures, data requirements and Actual Results documented?
- Is the test Documentation available to Clients?

Depending on the answers to these questions, the Testing Scope that is required will vary. Firms that are not comfortable with the quality of the vendor's testing will need to conduct more extensive testing, with multiple Test Phases.

Customization Testing

Depending on the type of Application and its expected use, it may be necessary to customize the Application prior to Implementation. The vendor will probably customize the Application if it is a Change to an existing Function, or the in-house staff could extend the Application if they develop a separate Module.

The Changes the vendor makes may be Client specific or may be added to the core Application in a release to the vendor's other Clients. In either case, the Function that is being developed will need to be extensively tested in the Planned Environment to make sure that the Changes work properly and meet the intended use.

Depending on how the Functional Requirements are documented and communicated to the vendor, there are opportunities for miscommunication on how the custom Code should perform, so testing in the Environment is essential to the successful Implementation of the customized Code as well as the vendor's core product.

Test Types for a Vendor Application

With the purchase or lease of a vendor product, it is important to tailor the testing effort to the Application. There are three main levels of Application Implementation, each with their own Test Types.

- Implement the Application "as is" with no customization
- Implement the Application with custom Code-Values or Rule-Sets
- Implement the Application with custom Programming Enhancements

The figure below is a Guideline for the Test Types that would be used to test each of the three Implementation levels:

TEST TYPES			
	IMPLEMENTATION WITH NO CHANGES	IMPLEMENTATION WITH USER DEFINED CODE-SETS OR RULE-SETS	IMPLEMENTATION WITH CHANGES
Business Review	Y	Y	Y
Technical Design Review	N	N	Y
Data Design Review	N	N	Y
Unit	Vendor	Vendor	Vendor
System Integration	Vendor	Vendor	Vendor
System	Y	Y	Y
Business Technical	N	Y	Y
Business Functional	N	N	Y
Business Integration	Y	Y	Y
Regression	N	N	N
User Acceptance	Y	Y	Y
Parallel	Y	Y	Y

Figure 12 – Testing of Vendor Applications

In each of the Implementation levels, the acquiring organization should expect that the vendor has performed Unit, System and Systems Technical Tests prior to delivering any Code for Implementation. However, the Test Team should validate this with the vendor, and if possible, inspect the results of the vendor's testing.

Authors' note: The best time to talk about whether or not the vendor will provide information regarding their testing experiences is before the contract is signed – not after.

Otherwise, the Test Team will have to conduct more extensive testing to validate the performance of the Application.

Category 2 - Testing a New Internally Developed Application

As with the purchase of a vendor Application, it is important to understand why a company decides to Develop a new Application internally. Some of the reasons are:

- New business requirements and no vendor Application exists
- New platform/infrastructure for an existing business to gain a competitive advantage
- Cost/performance of the existing Application is too expensive

Development of a new Application using in-house resources has advantages and disadvantages from a testing perspective.

- On the positive side, the Test Team is often able to participate in the process much earlier than with a vendor product, and this:
  - Makes it easier to prepare the Test Plans
  - Increases the level of Training of the Testers
  - Allows the Test Team to provide input during the Design Phase
- On the negative side, the Testing Scope must be more encompassing because all of the testing must be performed by the company.

Even with an in-house Development Project, an outside vendor often provides Programming resources when some specialized skills are needed for the

Development, or because an aggressive timeline for the Project requires more people than the organization can dedicate to the effort. In this case, the testing is the same as if all of the resources are internal staff since the management is in-house.

There are two sub-categories of internally Developed new Applications:
- Replace an existing Application
- Develop an entirely new Application

Analyzing Internal Applications

The Testing Scope required for either type of internally Developed Application is broader than for vendor-supplied Applications. To determine the Scope, the following topics should be reviewed.
- Completeness of Documentation
- Alternative sources of information
- Complexity of the Application
- Number of Interfaces required to Implement the Application
- Impact of the new Application on the existing Applications

Each of these factors needs to be explored in order to determine the level of testing that is needed for an acquired Application, and the factors are discussed in the following sections.

Source Documentation

When a new Application is actually an upgrade of a current Application, the Documentation that is needed to begin the Test Planning is usually more difficult to evaluate. There should be some Documentation for the current Application, but documentation normally is not maintained properly over the years, and the Functional Requirements Document for the new Application may be more Developer-oriented.

In this case, the people writing the Test Plan will have to review all of the available Documents. Some of the sources that could be used to support the Test Planning are:

Original Requirements for the Current Application

Depending on the Application's age, the original Business Requirements or Functional Specifications may be available. If the current Application has had many Changes through the years, the original documents were probably not thoroughly updated and may not be very useful but they will always serve as good reference material or as a starting point.

The Requirements for the new Application should include detailed information on all of the Application's Functions and how it will be used within the organization, including all automated Interfaces with existing Applications.

User Guides/Training Material

The current Application will usually have some type of User Guide and/or User Training material. These Documents also may not have been maintained, but will be helpful in determining the Application's Function, as well as how it is used in the organization along with the Code-Values and Rule-Sets that are in use.

Business Line Users

The End Users of the current Application can provide valuable input in the Test Planning process by providing guidance on the:

- Current Application
- Historical perspective on the decision to develop a new Application
- Reasons for choosing a new Environment and/or infrastructure
- Workflows and procedures once the Application is Implemented

Business Analysts

Often, the Business Analysts who defined the Requirements for the new Application can provide valuable information regarding not only the Functions and use of the Application, but also provide a historical perspective as to how decisions were made concerning the need for additional Functions.

Test Types for New Development

When a new Application is Developed in-house, all of the testing responsibilities are handled internally. The Test Types that are performed for these Projects differ slightly; however, the detailed testing will differ greatly both in how the Test Planning is conducted and how the testing is Executed. The main difference is that there is no Parallel Test for new Development for a new business since there is no current business or process to compare.

The following is a Guideline for the Test Types that could be used for the two main sub-categories for new in-house Development:

TEST TYPES	NEW APPLICATION FOR A NEW BUSINESS	REPLACEMENT APPLICATION FOR AN EXISTING BUSINESS
Business & Design Reviews	Y	Y
Data Design Review	Y	Y
Unit	Y	Y
System Integration	Y	Y
Systems Technical	Y	Y
Business Technical	Y	Y
Business Functional	Y	Y
Business Integration	Y	Y
Regression	N	N
User Acceptance	Y	Y
Parallel	N	Y

Figure 13 – Testing Required for New Applications

The Master Test Plan will focus on different testing activities depending on the reasons behind the decision to go to a new Application. Chapters VI, VII and VIII define the information required in the Master Test Plan, the Detailed Test Plans and the Project Plan.

Testing Changes to an Existing System

Testing Changes to an existing Application requires a very different approach than testing an entirely new Application. In this section we will discuss the requirements for testing Changes to internally Maintained Applications and for vendor Maintained Applications.

Category 3 - Testing Vendor Maintained Changes

If an organization has selected a vendor Application instead of Developing internally, the vendor will usually make the Changes that are requested and typically bundle them into a Software Release. In some rare cases, a firm will buy the source Code from a vendor and make their own Changes, and the testing for this is the same as for any other internal Code.

There are four sub-categories of Changes to an existing Application that is maintained by a vendor:

Enhancements

Enhancements are Changes to the Functions that were not provided in the original Application, and which will enhance the look, feel or function of the Application.

Regulatory Requirements

Regulatory Changes are required to keep the Application and its related processing and Reporting in compliance with the Regulatory Requirements that are dictated for the business. These Changes can be minor or extensive, and usually have strict externally imposed deadlines.

Technology Upgrades

In many organizations, the Functions of a current Application meet the businesses needs, but the platform or infrastructure needs to be upgraded to allow for expansion of the business or the Application, or to eliminate costly obsolete technology. The testing of this migration is often less difficult than for other types of Projects because the results can be compared to the current product, and the Parallel Test is the most effective Type of Test.

However, it is not always as simple as comparing results from the new Application to the current Application. Some types of upgrades require a more complex testing effort.

Web Applications

It may be that a current Application is being re-designed to be a web-based Application rather than mainframe-based. In this case, the Detailed Test Plans will have to focus on web-specific tests, such as:

- Browser options
- Edits and navigation using the User Interfaces
- Response time scenarios

It may be that the result of this type of Development will look very different while still performing the same Functions.

Database Migrations

There are business situations that may require that the current Application's Database(s) be migrated to a new structure or a new platform to meet the growing needs of the business or to support a technology migration to become a web-based product. In this case, the focus of the testing may be on exercising the Database's read and write Functions with less emphasis on the Application logic and User Interfaces (assuming that they are not changing).

Cost and Performance of the Existing System

Other considerations in starting the Test Planning process for the migration to a new platform are the cost and performance of the current Application. Very often, the key factor in the need to move to a new platform or infrastructure centers on either the Application's overall cost to run or its on-line and batch performance.

In this case, data should be gathered on any cost or performance aspects of the Application that are available. The Test Team should understand the areas of concern and the Service Level Agreements for the new Application so that they can focus the testing on those areas.

Defect Repair

Bug fixes are Changes that are made to repair Defects that were found in the current Application. These Defects are not always Code-related Issues. They may also be as a result of Design, specification or mapping discrepancies that are detected by either the Test Team or the End User. Some Issues that are initially thought to be Code-related may be ultimately identified as the result of incorrect File set-ups or incomplete Functional Requirements.

In each of these four sub-categories, the Test Process is essentially the same. The testing of the Changes has three main objectives:

- Ensure that the Change functions properly
- Ensure that the rest of the Application or process is not adversely affected by the Change
- Ensure that the Change does not negatively affect performance

Source Documentation

One difference with vendor Maintained Applications is that the Changes may be requested by a person in another company who is using the Application. For this reason, it is critical that the vendor provides Release Notes or some other Documentation that will describe the Changes in detail.

Depending on the vendor, the releases may be scheduled at regular intervals throughout the year, or as they are developed. Even though the releases could contain Changes from all of their Clients, each release must be installed because each release usually builds on the Code from the prior version.

Test Types

There are three sub-categories of Changes from vendors:

- Releases with requested Changes
- Releases with generic Changes that affect all companies using the Application
- Releases with specific Changes that only affect another company

For each sub-category, each Client will normally have to test all of the releases since any Changes requested by other Clients could affect the use of the Application. If the Change is a standalone Module requested by another company, the Module will not have to be tested, but the sending and receiving Applications should be tested to see if they are affected.

Therefore, for a release where the organization has requested Changes, more Functional Testing will be required, and for a release without requested Changes a Regression Test will likely be more important. A Regression Test will ensure that none of the Changes that were made for other Clients will negatively impact the use of the Application.

TEST TYPES	RELEASE WITH THE CHANGES	RELEASE WITHOUT THE CHANGES
Business Review	N	N
Technical Design Review	Y	N
Data Design Review	Y	N
Unit	N	N
System Integration	N	N
Systems Technical	N	N
Business Functional	Y	N
Business Integration	Y	Y
Business Technical	Y	Y
Regression	Y	Y
User Acceptance	Y	Y
Parallel	Possibly	Possibly

Figure 14 – Testing Vendor Changes

Category 4 - Testing Internally Maintained Applications

Testing Changes to internally Maintained Applications requires technical and political skills beyond those needed for testing vendor Applications. With internally Maintained Applications there are additional pressures, such as:

- Difficulty in criticizing peers (and friends)
- Internal managers often feel that the Application already works and the Developers know the business so testing can be perfunctory
- When internal Development is late, there is more pressure to reduce the testing Scope to meet the final delivery date

The Test Manager must overcome these obstacles, while managing the actual process of testing.

Sub-Categories of Changes

As with any other Application, the Changes to internally Developed Applications usually are:

- Enhancements
- Regulatory Updates
- Technology Upgrades
- Defect Repairs

Source Documentation

Access to the Documentation for Changes to an existing Application is just as important as for a new Application. The Development Team, Test Teams and End Users all need to fully understand the Changes and their impact on the rest of the Application. There are several potential sources of Documentation.

Business or Functional Requirements

The Requirements for an Enhancement or Regulatory Change should document any new Functions, Screens, Reports or Interfaces and any of the current Functions that are affected by the Changes. This should include all Code Changes, mapping Changes as well as Report layouts, User Interface Changes and Application Interface Changes.

Requirements from Regulatory Agencies

Where there are Changes that are made as a result of Regulatory Requirements, there will always be Documentation available from the governing body requiring the Changes. This Documentation, or an interpretation of the Requirements, is essential for both the Design of the Change and the related testing to ensure full compliance with the regulations.

Issues Tracking Report

When the Changes are made to correct a Defect in the Code, Design or mapping, the Development Team should have access to the Test Results to fully understand the Issue and to get the benefit of any examples of the Defect that were provided by the person or group reporting it. The Test Team also should have access to the person or group reporting the Defect in order to get additional details, if needed.

Test Types

To ensure that the Change is working properly, the Test Team needs to Execute many of the same Test Types that are performed on a new Application. The testing of the Change will include not only the Code that was modified, but also all other parts of the Application that could be affected by the Change. For instance, if a new Transaction Type is added to an accounting Application, the posting of the transaction would be tested, as well as the processing of the transactions through the Application and onto Reports or into Interfaces.

To test a Change to an existing Application, not all Test Types are required. The following figure is a Guideline for the Test Types that would be used for the four sub-categories of Changes for existing Applications:

TEST TYPES	ENHANCEMENT	REGULATORY CHANGE	TECHNICAL UPGRADES	DEFECT REPAIR
Business Review	Y	Y	N	N
Technical Design Review	Y	Y	Y	N
Data Design Review	Y	Y	N	N
Unit	Y	Y	Y	Y
Systems Integration	N	N	Y	N
Systems Technical	N	N	Y	N
Business Technical	Y	Y	Y	Y
Business Functional	Y	Y	Y	Y
Business Integration	Y	Y	Y	Y
Regression	Y	Y	Y	Y
User Acceptance	Y	Y	Y	Y
Parallel	Possibly	N	Y	N

Figure 15 – Testing Required for Changes

However, the main difference between testing a Change and testing a new Application is the process of Regression Testing. Regression Testing goes beyond testing the quality of the Change itself, and ensures that the rest of the Application is not affected by the Change that was made. For instance, if a new Transaction Type is added, the Regression Test will test the specific Transaction Type and test the other related Transaction Types to ensure that they still work properly. The Scope of the Regression Test is not usually as extensive as new Application testing because it is assumed that a Change to a Production Application implies that the original Application has been fully tested.

Also, unless the Change affects a substantial part of the Application, the Parallel Test does not need to be Executed. For most Changes, System Integration and System Technical (Stress and/or Performance) Testing is not necessary because the Changes do not normally alter the performance or structure of the Application. However, based on the size of the modifications, the Test Team, with input from the Development Team, may decide that a Stress or Performance Test is required to ensure that the Service Level remains at least as good as before the Change.

Category 5 - Conversion Testing

A Conversion could be required with a new Application or a Change in the business, or when the Application is acquired from a vendor or developed internally. In any case, the Code that Executes the Conversion should be tested in almost the same way as any other Development Project. A Conversion usually involves converting data from one Database to another, either in support of a new Application or because of a new platform or infrastructure.

Data Conversions

There are two basic types of data Conversions:

Data Mapping

In Data Mapping, the content of each Data Element in the source Application is directly mapped to a specific Data Element in the target Application. Once defined, this process is usually very straight forward, and much of the testing can be automated.

Data Transformation and Mapping

However, in most cases, all of the contents cannot be copied directly. Each Application may use different codes and abbreviations, and there could be other technical differences, such as the date structure. In this case, an intermediate table is required that is accessed during the Conversion, and the Conversion Program converts the data from:

- One format to another
- One Data Element to many
- Many Data Elements to one, or more

The testing for this process is more complicated since now the source and target information are different.

Source Documentation

The Documentation concerning the Conversion Test is critical to a successful migration of data from one Database to another. The Development Team, Test Teams and End Users all need to fully understand the Conversion and its impact on the rest of the Application and operations. The source Documents that are needed include:

Business or Functional Requirements

The Requirements for a Conversion will document all of the Functions, Reports, and Interfaces and any reconciliation that is needed to support the data Conversion process, and will contain the same level of information as can be found in any Business or Functional Requirement Document.

Source Database

To understand the Source Database the Test Team should have the following Documents.
- Database Design or File Layouts
- Data Dictionary or Data Element Definitions
- Data Flow Diagrams

Target Database

To understand the Target Database the Test Team should have the following Documents.
- Database Design or File Layouts
- Data Dictionary or Data Element Definitions
- Data Flow Diagrams

Conversion Processing

For Conversions involving large Databases and complex data, the Conversion may also have a few specific processes:

- Static data Conversions
- Dynamic data Conversions
- Conversion workstation to Control the process
- Conversion reconciliation

The Scope of Conversion Testing should be treated the same as any other new Application with one difference; the testing must validate that the converted data is available for use by the new Application from all aspects of the operational processes. The objective of Conversion Testing is to:

- Ensure that any automation Developed to support the Conversion will work correctly
- Ensure that the data is correctly converted
- Ensure that all of the converted data can be used by the Application throughout the entire process

Test Types

The testing that is performed on the Conversion process is usually limited to the following types:

TEST TYPES	CONVERSIONS
Business Review	N
Technical Design Review	N
Data Design Review	Y
Unit	Y
System Integration	N
Systems Technical	N
Business Functional	Y
Business Integration	Y
Business Technical	Y
Regression	N
User Acceptance	N
Parallel	N

Figure 16 – Testing for Conversions

The Systems Technical Tests, including the Stress Test, usually can be eliminated because the Conversion is normally an isolated event and there are no on-going performance Issues to address in the testing, unless:

- A large Database needs to be converted in a restricted amount of time
- The new Database has not been Stress Tested under high-volume Conditions

There is no need for a UAT, and the Parallel Test does not need to be performed because there is no parallel process for comparison. The Functional Test will cover each of the areas on the Conversion process individually. The Business Integration Test should ensure that the converted data also could be used by the remaining test activities.

Summary

Based on the Category of Testing Project, the first step in Test Planning is to decide on the Test Types that will be needed to fully test a new Application, Changes to an existing Application or a Conversion. The categories and sub-categories are:

- New Application
    - Category 1 - Buy from a Vendor
        - No Change to the Vendor Application
        - Implement Rule-Values and Code-Set Changes
        - With customization
    - Category 2 - Build In-House
        - Rebuild an Existing Application
        - Build a New Application
- Changes
    - Category 3 - Buy from a Vendor
        - Release Contains Company-Specific Changes
        - General Release Does Not Contain Company-Specific Changes
    - Category 4 - Build In-House
        - Enhance
        - Regulatory Updates
        - Technology Upgrades
        - Defect Repair
- Category 5 - Conversions

Variables	
AB	Available Budget
AC	Average Cost/Person
ER	Effort Required (Total)
PR	People Required (Total)
TA	Time Available (Total)
TR	Time Required/Task
RB	Required Budget

Determining the effort required and the effort per Task is based upon:
- Complexity of the Project
- Length of the Project
- History with the Application and the team
- Skills of existing resources

This chapter identified the Categories of Testing Project that are typically used for each type of Project. The next step in the Planning process is to prepare to determine the testing methodology, which is discussed in Chapter V – Getting Started.

V. GETTING STARTED

The major Tasks of Planning, Executing, Controlling and Closing are activities that are necessary in all of the Testing Project Categories that were discussed in Chapter IV, and are the essential Components in managing:

- A full Project
- A Phase of a Project
- A Task within a Phase of a Project

The initial Test Planning and the groundwork for the Master Test Plan should begin during the Design Phase of the Systems Development Life Cycle once the Requirements for the Application have been established. Although it is preferable for the Test Manager to be in place at the start of the Project, it is critical for him/her to be in place during the Design Phase to begin the Planning. This is when decisions are made that will shape how, when and where the entire test effort is conducted.

If an experienced Test Manager is not in place to participate in this initial phase, decisions might be made that could have a negative impact on the quality and timeliness of the overall deliverable. Some of the critical decisions made at this time impact the length of the test, the number of people needed, and the number of Test Regions required. The Test Manager must be in place to accurately assess these Requirements, and to structure the remaining aspects of the test.

> **Authors' note:** The Test Manager has to become a part of the Project Team. This is easier if he/she is on board from the beginning.

To Implement the testing process efficiently, Testers need to understand why software has Defects. In general, Defects (errors, bugs, etc.) result from the following:

- A lack of communication, or miscommunication between the people who know what they want the Application to do and the people who are responsible for building it.
- The overall complexity of modern software makes it unlikely that any one person will be able to understand all facets of the Application. The more Developers working on the Code, the greater the chance for Defects.

> **Authors' note:** Defects result from more than just Programming Errors.

- The complexity of modern Code and Development techniques also increases the level of education and experience that is required to use the available Development tool(s).
- No one intends to do a bad job, but people make mistakes. Developers do not set out to build bad Code that will be delivered late and over Budget, but some bad Code is inevitably delivered in complex Projects.
- There might not be a structured process in place, such as a SDLC.
- Requirements that are defined today may be influenced by events tomorrow and may no longer be valid. This doesn't always get translated into a Document that is given to the Developers.
- The pace of change in business increases the pressure to Develop Code quickly. This haste creates a situation where the Code is considered "good enough" and Defects are accepted.
- People want to be heroes and sometimes agree to do more than they can deliver.

- People are often unwilling to admit that they do not understand something.
- If the Code being enhanced is not properly documented, the Developer may create Defects that meet the Requirements of the erroneously defined Code.
- Some firms still have insufficient Development tools, Guidelines and Standards, and make it difficult for Developers to do a quality job.
- People may not be given enough time to learn how to properly use new tools or how to work with a new product.

> **Authors' note:** By managing the many points that cause Defects, you can have a successful Project.

With all of these potential sources of Defects, there is a need to test. Although organizations may differ in how they test, in general every organization typically goes through the high level processes of Planning, Execution, Controlling and Closing. The following chart identifies the processes and the major testing activities that are the key Components of a successful test.

\multicolumn{3}{c}{MAJOR TESTING ACTIVITIES}		
PROCESS	**ACTIVITY**	**CHAPTER**
PLANNING	Understand the Testing Process	Chapter II, III
	Determine the Category of Testing Project	Chapter IV
	Decide on the Testing Methodology	Chapter V
	Master Plan	Chapter VI
	Detail Plan	Chapter VII
	Project Plan	Chapter VIII
EXECUTING	Prepare the Tests	Chapter IX, X
	Execute the Tests	Chapter IX, X
CONTROLLING	Manage the Tests	Chapter XI
	Managing Cost and Risk	Chapter XII
	Automate the Tests	Chapter XIII
CLOSING	Assess Application Readiness	Chapter XIV
	Obtain Sign-offs	Chapter XIV
	Implement	Chapter XIV
	Conduct the Post Project Review	Chapter XIV

Figure 17 – Major Testing Activities

Selecting the Testing Methodology

One of the first things a Test Manager must do in the Planning process is to determine whether the organization has Policies, Guidelines or Standards that must be followed. Although the Test Manager's primary concern is whether or not there are company requirements that directly impact testing, there are other outside conventions that could affect how the test is Planned, Executed and Controlled.

> **Authors' note:**
> Project Management is what gets done.
> Quality Management is how it gets done.

Many of these methodologies fall under the general categories of Quality Management and/or Project Management. Ideally, a Project is best managed within a framework that integrates the relevant guiding principles of Quality and Project Management Programs, along with a comprehensive Test Management strategy.

The following section provides a review of the activities typically governed by each of these disciplines. Although the Implementation of these disciplines may vary from organization to organization, the key Components remain fairly consistent.

Quality Management

As discussed in Chapter I, Quality Management supports the overall quality objectives of an organization that can be defined for the Application at the organization level or Project level, as well as for the management of the Project Implementing the Application. This definition can include, but is not necessarily limited to the Guidelines and Industry Standards[9] that have been established.

CATEGORY	STANDARDS
Documentation	Project Management Documentation
	Operations Evaluation and Acceptance
	Client Interactions
Process	Project Management Guidelines
	Software Development Life Cycle
	Software Engineering Standards
	Operations Procedures
	Audit Requirements
Quality Metrics	Timeliness
	Productivity
	Error Rates
Vendor Management	Service Level Agreement
	Payment Schedule

Figure 18 - Categories of Standards

If a Quality Program does not exist or is not enforced for the business, the Test Manager should develop one, or ensure that it is used.

Project Management

Project Management has evolved into a methodology that utilizes a range of tools and processes to manage Projects efficiently and effectively. The conventions that have been established generally define:

1. The elements of the Project to be tracked and managed, such as:
   - Tasks
   - Resources
   - Timeframes
   - Dependencies
   - Quality
   - Cost
   - Risk
   - Communications

---

[9] The various Industry Standards that are relevant for testing are discussed later in this Chapter.

2. The processes to be managed, such as:
   - Planning
   - Executing
   - Controlling
   - Closing
3. The tools that are needed to manage a Project, which include:
   - Project Plans
   - Test Plans
   - MIS

Based upon the authors' experience, the keys to successfully implementing a Testing Strategy are:

KEY	DEFINITION
Requirements	Business and Functional Requirements must include clear, complete, detailed, cohesive, attainable, testable Requirements that are agreed to by all players.
Standards	The Project Manager should ensure that the Test Team is using the most current and relevant Guidelines, Standards and Conventions that are available.
Best Practices	The entire Test Team should be willing to learn from the experiences of others and the Best Practices that have evolved in the Financial Service Industry.
Realistic Timeframes	The Test Project Plan must allow adequate time for Planning, Training, Design, Testing, Defect Repair, Re-testing, Changes, and Documentation. The Plan should not rely on overtime and seven-day workweeks, and lead-time for hardware, software and premises should be adequately considered.
Skilled Staff	The staff who are assigned to work on a testing project should have a broad set of experience and training, and be very flexible in order to respond to the constant shifts in direction that are typical of a Testing Project.
Thorough Testing	The Project Manager must start testing early in the Project, and Re-test Repairs or Changes quickly. The Plan should consider that not all of the Code corrections will work when they are received.
Avoid Scope Creep	The Project Manager must be prepared to resist Changes and additions to the Scope once Development has begun, and be prepared to explain the consequences of the Changes on the Application, the Test Project Plan and the Budget.
	If Changes are necessary, the impact on the Plan should be clearly defined and communicated. While the Test Project Plan should include some time for unforeseen Changes, in general, significant Changes cannot be absorbed without affecting the timeline, the Budget or the quality of the testing.
Automation	The Test Manager should review the available test automation tools and be willing to consider any source of automation in order to control the project more effectively.
Communication	The Project Manager should require continuous communication among all of the members of the Project Team. This should include communication that occurs via: • Walkthroughs and inspections • Group communication tools (e-mail, groupware, etc.) • Networked Defect-tracking tools • Change management tools • Accessible, current electronic Documentation The Project Manager should ensure that influential people and interested managers outside of the Project Team are updated on the Project's process, or on any delays.

Figure 19 – Successful Project Management Criteria

Test Management Conventions

Detailed testing conventions can also exist within a firm to supplement the high level Guidelines that are established for the overall management of a Project. These conventions typically define:

- Type of Test

- Scope
- Timeframes
- Dependencies
* Format and Content
  - Test Plans
  - Test Calendars
  - Business Scenarios
  - Test Cases
  - Test Procedures
    - Objectives
    - Test Transactions
    - Expected Results
    - Test Log with Actual Results
  - Issues Tracking Reports

> **Authors' note:** Whether you use existing Standards or create your own Guidelines, you must define a process in order to effectively manage a Project.

If Standards, Guidelines or Conventions for any of these activities exist, the Test Manager should use them to define the Test Types that must be conducted, the Test Documentation that is required, and an overall framework for managing the testing process. If test Standards do not exist, the Test Manager must either create the relevant testing controls for each of these activities, or integrate existing general Standards into their overall Test Management strategy.[10]

The following section provides information on some of the key national and international organizations that are responsible for defining general Standards and/or Guidelines for software Development Projects, and a summary of those Standards that could be of interest to the Test Manager.

> "Facts do not cease to exist because they are ignored."
> Aldous Huxley

Standards Organizations

There are many organizations, both nationally and internationally, that have established Standards in support of software engineering activities. Several of these key organizations are described in detail in this chapter. The SESC[11] (a committee within the IEEE[12], which is one of the primary Standards Developers in the US) and the ISO[13] (one of the primary international Standards organizations), produce many of the Standards that are utilized today in software Development initiatives. These organizations, along with a number of others, have recognized the need to put processes in place to integrate their own Standards, and to align with and integrate the Standards established by other organizations where there are anticipated synergies.

---

[10] The authors provide their own Standard definitions and processes throughout this book, which have been derived from the SESC and experience in the Financial Services Industry.

[11] Software Engineering Standards Committee (SESC)

[12] Institute of Electrical and Electronics Engineers (IEEE)

[13] International Organization for Standards (ISO)

The SESC is one of the organizations that have taken steps to revamp their own Standards and to integrate outside Standards into their Guidelines. As a result, this collection is a relatively comprehensive reference source. The following section provides information about the SESC and a few of the other key outside organizations whose Standards and Guidelines have been integrated into the SESC suite of Standards.

This is not a complete list of all of the Standards organizations; however, it does provide an overview of some of the key organizations and Standards of interest to IT professionals, including the Test Manager. Moore[14] has published a useful book with a comprehensive listing of the Standards organizations along with a summary of some of the most widely used Standards, and Schmidt[15] has written a guide for Software Engineers that addresses Developer testing.

- Moore's book, *Software Engineering Standards: A User's Road Map* was written in cooperation with the IEEE/SESC and with the US Technical Advisory Group (TAG) to the ISO/IEC. The International Organization for Standards (ISO), and the International Electrotechnical Commission (IEC), jointly created this subcommittee to work on the integration of their Standards. This book provides information on the key Standards organizations and the Standards that support software engineering activities, and provides a map which describes what they are, how they relate, and how to choose those that will work best for an organization.

  Authors' note: Several books are available that discuss Industry Standards in depth.

- Schmidt's book, *Implementing the IEEE Software Engineering Standards*, has been endorsed by the IEEE Standards Information Network, and has been recognized by the SESC as a useful guide for software practitioners applying software engineering Standards. This book provides a practical guide for phasing in the most important principles identified by the IEEE SESC according to a maturity model developed by Schmidt. In a simplified approach, this model helps Development Managers determine the maturity level of the organization, and how to apply the Standards accordingly.

Institute of Electrical and Electronics Engineers, Inc., (IEEE) – Software Engineering Standards Committee (SESC)

The SESC is a committee within the Computer Society of the IEEE that produces Standards in support of software engineering Best Practices, and is one of the primary proponents of Standards in the United States. There are approximately 50 Standards that exist, covering a broad range of software engineering activities.

Much thought, time and effort has gone into the creation of these Standards, and they contain a wealth of information that can help the IT professional, including the Test Manager. However, since they were not originally written to work as an integrated set of Guidelines, it is difficult to know which ones to use, when to use them, and what the differences are between them. It is the Test Manager's responsibility to research, evaluate, incorporate and or establish separately the

---

[14] James W. Moore, *Software Engineering Standards: A User's Road Map*, IEEE Computer Society Press, Los Alamitos, California, 1997.

[15] Michael E. C. Schmidt, *Implementing the IEEE Software Engineering Standards*, SAMS Publishing, Indianapolis, Indiana, 2000.

right combination of Quality Management, Project Management and Test Management Standards. To assist in this process, there are supplemental guides that have been developed to help evaluate the myriad of Standards.

The SESC has grouped the IEEE Standards into several categories that appear in Moore's book as follows:

- Project Management
- Planning
- Development Life Cycle Processes
- Individual Processes
- Documentation
- Measurement
- Tools
- Re-use
- Other

Imbedded within many of these high level categories are Standards that are helpful to the Test Manager.

CATEGORY		IEEE STANDARD
Planning	730-1998	Software Quality Assurance Plans
Project Management	10444-1993	Classification for Software Anomalies
Documentation	829-1998	Software Test Documentation

Figure 20 - IEEE Testing Standards

A full listing of these Standards and their groupings are contained in Moore's book, and both Moore's and Schmidt's books should help Testers understand the Standards, and how they can be integrated into a Project.

The International Organization for Standards (ISO)

The ISO is a worldwide federation that is comprised of numerous national Standards bodies and is one of the most widely referenced international Standards organizations. One of the primary Standards established by the ISO that has been incorporated for use in software Development initiatives is the ISO9000 family of Standards. The Standards in this grouping facilitate the establishment, Implementation and Maintenance of an effective Quality Management process.

A number of the ISO Standards have been adopted by the SESC as key IEEE Standards for Quality Management and have been integrated with their own existing Standards.

Project Management Institute (PMI)

PMI is an organization that has written Guidelines for Project Management, and has organized a program that certifies Project Managers as Project Management Professionals (PMP). The PMI Certification Program Department received ISO 9001 recognition, and their book the *Guide to the Project Management Body of Knowledge (*PMBOK Guide) which contains generally accepted Project management practices, has been adopted by the SESC as an IEEE Standard for Project Managers to supplement their own Project management Standards.

Software Engineering Institute – Capability Maturity Model

The Software Engineering Institute (SEI) was instituted at Carnegie-Mellon University by the U.S. Defense Department to help improve the Software Development processes. The Capability Maturity Model (CMM) was established by the SEI. The model consists of five levels of organizational 'maturity' that define a company's effectiveness in delivering quality software.

Authors' note: The CMM is widely used as a QA tool by large organizations.

The model is geared to large organizations such as large U.S. Defense Department contractors; however, many of the QA processes involved are appropriate to any size organization, and if applied thoughtfully can be helpful. The median size of the organizations that have been tested has consisted of about 100 Developers.

Organizations can receive CMM ratings by undergoing assessments by qualified CMM Auditors. The levels, as defined by SEI are:

Level 1

Level 1 is characterized by chaos and periodic panics, where heroic efforts are required of individuals to successfully complete Projects. Few if any processes are in place; successes may not be repeatable.

Level 2

Level 2 consists of software Project tracking, Requirements management, realistic Planning, and configuration management processes; and successful practices can be repeated.

Level 3

Level 3 firms utilize Standard Software Development and Maintenance processes that are integrated throughout the organization, a Software Engineering Process Group is in place to oversee software processes, and Training courses are used to ensure understanding and compliance.

Level 4

In Level 4, metrics are used to track productivity, processes, and products. Project performance is predictable, and quality is consistently high.

Level 5

In Level 5, the focus is on continuous process improvement. The impact of new processes and technologies can be predicted and effectively Implemented when required.

The CCM concept is based on the fact that the process of Implementing quality assurance evolves over time:

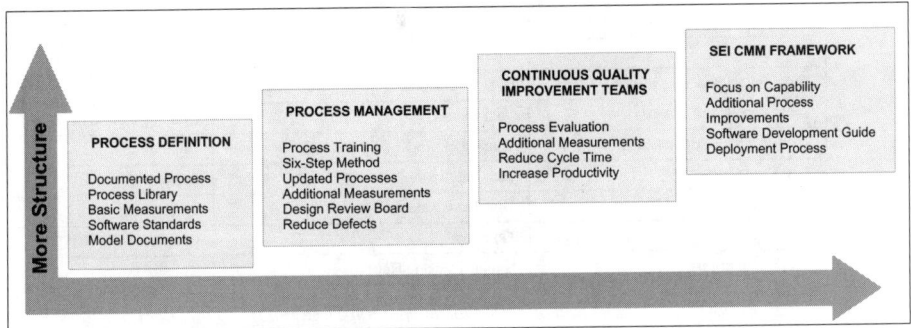

Figure 21 - Process Improvement Over Time [16]

The ratings for all of the organizations that are reviewed by CMM certified Auditors are reported to SEI, so some very good statistics are available. The percent of firms that have attained a specific level are shown in the following figure.

	1992 - 1996	1997 - 2001
Level 1	67%	27%
Level 2	23%	39%
Level 3	13%	23%
Level 4	2%	6%
Level 5	.4%	5%

Figure 22 – Firms Attaining CMM Level 1 to 5 Over Time [17]

Documentation Standards

It is very easy to get bogged down in testing paperwork. To be efficient, the Testers should spend the majority of their time actually testing and not on completing the many forms that are required to manage the Test Process. The Documents that are actually used to manage the tests should be carefully selected, based upon a number of factors, including:

- Experience level of the Test Team
- Probability that the test Documents will be re-used and by whom
- Reporting requirements from the Test Manager and other interested parties
- Relationship between the Test Team and the other interested parties

As previously mentioned, ANSI Standard 829 is one of the many Standards that could be of interest to the Test Manager. This particular Standard identifies a number of different Documents that should be used in the testing effort. Some of these are:

---

[16] George Yamamura, Gary B. Wigle, SEI CMM Level 5: For the Right Reasons, *Boeing Defense and Space Group, 2000*

[17] Software Engineering Institute

ANSI 829 TERMINOLOGY	TERMINOLOGY USED IN THIS BOOK	DOCUMENT COVERED IN:
Test Plan	Master Test Plan	Chapter VI
Test Design Specification	Business Scenario	Chapter IX
Test Case Specification	Test Case	Chapter VII and IX
Test Procedure Specification	Test Procedure	Chapter IX
Test Procedure	Test Procedure	Chapter IX
Test Item Transmittal Report	Test Transaction	Chapter IX
Test Log	Test Log	Chapter IX
Test Summary Report	Issues Tracking Report	Chapter IX

Figure 23 - Comparison of ANSI Documentation Terminology to this Book

You can get more detail on what is intended by each of these terms by referencing the complete ANSI Standard 829. While the Test Manager should make the decision as to the content and format of the test Documents, some of the following topics should be included at a minimum:

- Master Test Plan
- Business Scenarios
- Test Case
- Test Procedure
- Issues Tracking Report

Some of the other information recommended in ANSI Standard 829 can be added to these Documents, depending upon what is needed from a reporting, Training and status standpoint.

Software Testing Best Practices [18]

In addition to Standards, there are some Guidelines that can be considered. The concept of Best Practices is one that allows professionals to learn from the experiences of others. IBM has conducted considerable research in the area of software testing and identified "twenty-eight Best Practices that contribute to improved software testing." These practices are not necessarily related to test automation tools, but they are fundamental management tools for the testing process,

> **Authors' note:** Code can be Defect-free, and still not meet the End Users' Requirements. This is just as much a failure as if the software has pervasive Defects.

and if they are considered while automation tools are selected and implemented, the Test Manager will have a greater chance for success.

IBM grouped the Best Practices into three areas: Basics, Foundational, and Incremental. Seventeen of these practices are relevant to our discussion and are presented here, and have been integrated into the process we discuss throughout the book. Not every one of these practices will be appropriate for every firm, and the Test Manager should make a decision to use or not use them.

---

[18] Ram Chillarege, IBM Research - Technical Report RC 21457 Log 96856 4/26/99

Best Practices - Basics

Functional Specifications

As discussed in the section on the SDLC, one fundamental step in developing any Project is the Functional Requirements phase. While it is important from a Software Development perspective, it is equally important for testing. Since the Functional Requirement defines what it is that the users want, it is this Document that the Testers must use to determine if the Code that is delivered meets the users' Requirements.

Testers must write Test Cases that will validate that the Code meets the defined Requirements. Test Cases based upon the Functional Requirements are generally used during Black Box Testing, and these tests can be created as the Code is written.

Reviews, Inspections and Walkthroughs

The concept of Software Inspections was first defined in the 1970's and has been used intermittently since then. Most often used by large firms, Software Inspections can be an effective technique to identify and eliminate Defects from software.

The Software Engineering Institute defines Software Inspection as where "a small peer group of software Developers systematically reads a Document, identifying and classifying (potential) Defects as they are encountered. Participants play specific roles in this process, which is intended to improve Application quality, and to make the software production process more visible and more controllable."[19] Therefore, the inspectors do not look at the Code. They review a Requirements Document or a Test Plan to look for logical gaps and potential inconsistencies, before the Coding begins.

SEI goes on to point out that Software Inspection is similar to other types of software technical reviews, such as Walkthroughs, since software is examined and discussed by a group of software Developers and Business Analysts in a very structured and formal way. "Inspections are characterized by:

- Explicit entry and exit criteria
- Individual preparation by inspectors
- Defined roles of moderator, reader, producer, and recorder; Training for moderators
- Use of a checklist
- Limitation of discussion to identification and classification of Defects
- A requirement that successful completion of rework is necessary to complete the inspection
- Formal data collection, Reporting, and analysis"[20]

Formal Entry and Exit Criteria

The overall process of Software Development is based upon an Input-Process-Output model. The testing concept using this model requires that the inputs

---

[19] L. E. Deimel, Scenes of Software Inspection, Software Engineering Institute, CMU-SEI-91-EM-5
[20] ibid

and outputs be formally defined at each step in the Application so that they can be checked outside of the Code.

Functional Test - Variations

Most Functional Tests are written as Black Box tests based on the Functional Requirements Document. Developing the variations of Functional Tests that must be performed requires the Test Manager to define a series of input Conditions that are tested and compared to the Expected Results. The goal in writing this range of tests is to establish a repeatable model that will provide the widest coverage with the least effort.

Multi-Platform Testing

The Test Manager must understand the complete range of platforms that the Application can be run on. This is particularly important in the current Development Environment where, for example, many different types of browsers can access a single web-based Application. If so, the Test Process must allow for testing the Application on all of the different platforms.

Internal Betas

Once an Application has been tested, it could be given to a limited number of End Users (or selected Clients) to use the Application and see if any new Defects are identified, or if modifications are required to the User Interface.

Automated Test Execution

Automated Test Execution should be considered when the volume of Test Cases justifies the effort and cost of automation. Automated Execution should also include automatically generated logs of successes and failures, with sufficient information for the Developer to find the source of the Defect.

Nightly Builds

The concept of frequent (or nightly) build of Code requires the following:

- Code is added to the Application continuously as it is released by the Developers
- Automated Regression Testing is performed continuously in the background
- Code for new Functions must be tested immediately

Best Practices - Foundational

User Scenarios

Originally, each Application had its own User Interface. Increasingly, it is popular to Develop a Standard User Interface that accesses and works with all, or many, of the other user Applications. This simplifies User Training, but it complicates the testing effort. In order to test this common UI, a complete Test Library must be maintained (and updated) so complete Regression Tests can be performed every time a Change is made to any of the underlying Applications.

Usability Testing

Usability Testing is typically performed after an Application has passed all of the QA tests. In this Best Practice, the concept is to let the users evaluate the ease of use of the Application and provide suggestions to the Developers.

Automated Test Generation

It is generally accepted that writing the Test Cases uses up to 30% of the overall testing Budget, so this is clearly an area that should be examined for automation. In general, the available testing tools historically produced too large a set of Test Cases and thereby reduce the overall efficiency of the testing effort; however, this area should be continuously monitored since once a solution is found, it could be a significant cost and time-saver.

Best Practices - Incremental

Teaming Testers with Developers

Teaming Developers with Testers has proven to be a successful way to improve the Test Cases and the Code. For example, at Microsoft, every Developer is teamed with a Tester so that the Tester learns what the Developer is doing and the Developer knows that his/her Code will be examined in detail. This has to be managed to avoid either an adversary relationship or a situation where the Tester becomes too close to the Developer and too protective.

Code Coverage

There are several types of automated measurements that can be made to determine how much of the Code has been exercised during testing. This should include the Code statements, branches, and data.

Automated Environment Generator

Creating an automated Test Environment to Execute Test Cases in multiple Environments can take a significant amount of time that is based on the number of operating Applications, versions, and Code that runs on multiple platforms. Some tools exist that can mimic multiple Environments to simplify this process.

Memory Resource Failure Simulation

Although Developers do not need to be as concerned about memory utilization as they did when computers were far more limited, they still need to be concerned about whether PCs will have to be upgraded in order to utilize the new Application. For instance, many C Programs in Unix Applications have not been written efficiently and consequently use too much memory. Tools exist that help Testers determine whether or not the Program was Designed and Coded efficiently.

Minimizing Regression Test Cases

This Best Practice recognizes that many organizations have a large number of legacy Applications and many Test Libraries of regression transactions. Very often, there is considerable overlap among these Test Libraries, and there can be gaps. This Best Practice encourages Testers to evaluate their Test Libraries and establish a minimal set of Test Cases and Test Transactions.

Bug Bounties

Many years ago IBM established a monetary 'bounty' for Testers (internal and external) who found Defects. Microsoft has also used this technique, and these firms have found that, on average, a larger number of Defects are identified for Programs that are tested with a bounty. While more Defects are found, the resulting Applications have a higher degree of quality and reliability.

Sizing the Project

In order to establish the initial staff levels, the Project Management Team must first estimate the size of the Project, which at a very high level can be subjectively classified as Small, Medium or Large. However, a systematic three-step approach can be taken to determine the size of the Project:

- Identify the relevant variables
- Assign a weight to each variable
- Calculate the weighted total for the Project

In the following figure we have assumed that we have already identified the numerous detailed variables that are relevant to this Project and have grouped them into the seven summary topics[21] that are shown. The details have been aggregated and the total for each topic is shown, along with a weighting factor.

	SCALE						TOTAL	VARIABLE WEIGHT	WEIGHTED TOTAL
	<<<<	1	2	3	4	5	>>>>		
1. Single Application Development		X					Multiple Application Development	1	5	5
2. Simple Logic						X	Complex Logic	5	5	25
3. Known Application Type				X			New Application Type	3	3	9
4. No Interfaces		X					Many Interfaces	1	4	4
5. Single User Organization						X	Multiple User Organizations	5	3	15
6. Limited Business Impact				X			Significant Business Impact	3	5	15
7. Sufficient Internal Staff Available				X			Few Internal Staff Available	3	2	6
TOTAL										79

Figure 24 – Sample Project Weights

Given the Weighted Total of 79 for this Project, based upon our experience, this new Application Development would be considered a Medium Size Project, as shown in the following chart.

IDENTIFY THE PROJECT CLASSIFICATION			
	Between	Between	Over
If the weighted score is:	1 - 50	51 -100	100
Then the Project is classified as:	Small	Medium	Large

---

[21] The details behind this process are available from the authors.

			X	
New Application Development			X	
Purchased Package				
Changes to Existing Products				

Figure 25 – TSG Project Classification Ranges

The actual decisions on how to organize and staff a Project should be made by the Test Manager, based upon the specifics of the Project, the Test Manager's experience, and the following Guidelines.

PROJECT SIZE	STAFFING GUIDELINES
Large	Frequently one or more people are assigned to each Function or Sub-Function
Medium	More than one Function may occasionally be assigned to one person
Small	Typically, more than one Function is assigned to one person

Figure 26 - General Staffing Guidelines

As a Medium Size Project, this example would not require that all of the activities be individually staffed. For example, there may only be one Testing Section Manager, and it may be that the individual Developers will perform the Unit Tests. The following grid illustrates the testing activities that might be combined for this Project:

TEST TYPES AND FUNCTIONS [22]	RESOURCES				
	DEVELOPERS (MULTIPLE)	DATABASE MANAGER	TEST MANAGER	TESTERS (MULTIPLE)	END USERS
Business Analysis					
• Concept and Analysts					X
• Business Requirements					X
• Functional Requirements	X				X
• Business Review	X		X		X
Design	X				
• Technical Design Review	X				X
• Data Design Review	X	X	X		X
Programming	X				
• Unit Test	X				
• Systems Integration Test	X				
• Systems Technical Test	X				
– Stress Test	X			X	
Test Team Testing					
• Business Technical Test				X	
• Business Functional Test				X	X
• Business Integration Test				X	X
• Regression Test				X	X

---

[22] The various Test Types are discussed in detail in Chapter III.

End User Testing					
• User Acceptance Testing					X
• Parallel Test				X	X
Database Administration		X			
Test Section Management			X		
Tools Administration			X		

Figure 27 – Sample Resource Requirements

In a Large Size Project, each of the activities may be assigned to unique resources or teams of people for each activity. For example, in the case of a Securities Lending Application that Interfaces with a Custody Application, one team may be dedicated to testing the Securities Lending Functions, while another team is dedicated to testing the Interfaces with the Custody System.

Summary

Another reason that Guidelines and Standards are necessary is that while every organization says that they only hire the best people, by definition, almost every organization must have some people who are not "the best." Despite this inevitable range of talent, we have never found people who say that their goal is to deliver poor, late or unusable Code. People want to do well; some are naturally better than others; however, everyone can do better with the right Guidelines, Standards, Tools and Training.

Once the foundation for the Quality and Project Management methodologies have been established, the Test Manager must focus on preparing for the test effort itself. Understanding the test Standards if they exist, or establishing them if they do not exist, is the first step in a multi-step process. Typically the published Industry Test Standards do a good job of defining test terminology, Test Phases and types of Documentation, but they do not provide guidance on how to establish an action Plan using that framework. This is the purpose of this book.

Chapters VI through XIV describe the remaining test activities from Planning through Closing. This information is presented from a practical standpoint, based upon the authors' hands-on experience, so that a Test Manager can successfully Plan and Execute a test that best suits a specific organization while conforming to the company's requirements, as well as formal and informal industry Guidelines and Standards.

The other key points in this chapter are:

- Test Planning should start during the Design Phase of the Project
- Defects will occur and testing is needed to detect them
- Several types of Standards are available to help the Test Manager establish the Plan and approach
- Testing must be managed and Standards and Best Practices exist to aid Test Managers
- The size of the Project can be estimated quantitatively
- Small, Medium and Large Projects should be staffed differently

VI. MASTER TEST PLAN

After determining the size of the Project, the Test Project Category and the Test Types that are required, the next step is to create the Master Test Plan. Writing the Master Test Plan helps the Test Manager define, organize and manage the testing effort.

Once established, the Master Test Plan will be used as the basis for conducting and managing virtually all of the testing activities. Therefore, it is important to establish a realistic Master Test Plan that defines what is being tested and what is not, as well as when the tests will be conducted and completed. A Test Manager should not establish a Master Test Plan that he/she never expects to fully execute.

Authors' note: The Master Test Plan must be realistic, and the Test Manager should be held accountable.

Master Test Plan Objectives

The objectives of a good Master Test Plan are:

Document the Test Scope

The Test Team should consider the entire testing process as they write the Master Test Plan since the details will flow from general topics to very specific details for every phase of the Project. The information that is established as part of the Master Test Plan will allow the Test Team to efficiently create Test Cases that provide a full coverage of the Functions and the range of Conditions.

Provide Direction to the Test Team

The Master Test Plan provides the Test Team with a structured approach to testing so that each Tester will understand their role in the process, and the team leaders will be able to manage the process and update their managers on the Project's progress.

Provide Information to Interested Parties

There are many parties that are interested in the test Planning process, as well as Test Execution. Therefore, the Master Test Plan should provide enough information so that all of the participants in the Project and all other interested parties can become comfortable with the testing effort as a whole.

Master Test Plan Standards

There are many Guidelines and Standards that are relevant when writing a Master Test Plan. One of the more widely referenced Standards, ANSI/IEEE Standard 829-1983 [23], defines the Master Test Plan as including the Scope, approach and schedule for the testing. The Master Test Plan, according to this Standard should have the following sections:

- Test Plan Identifier
- Introduction
- Test Items
- Features To Be Tested
- Features Not To Be Tested
- Approach

---

[23] This was presented in Chapter V.

- Item Pass/Fail Criteria
- Suspension Criteria and Resumption Requirements
- Test Deliverables
- Test Tasks
- Environmental Needs
- Staffing and Training Needs
- Schedule
- Risks and Contingencies
- Approvals

> **Authors' note:** Customize the Master Test Plan to meet the Project's Requirements, not to conform to an arbitrary Standard.

In the authors' experience, ANSI's approach to the Master Test Plan does not always meet the needs of every organization. In the course of our work with many different financial services firms, we have established a Master Test Plan approach that has worked well for the diversified Testing Projects that we have managed and supported.

However, since each organization needs to establish Master Test Plan Guidelines that can be used as the basis for its own Testing Projects, Test Managers should feel free to adapt our proposed Master Test Plan to their unique requirements. Regardless of the approach selected it is important that one base Document is used by all of the participants to ensure that all of the key points of interest to the Test Team, Management, End Users, Auditing, etc., are covered.

Master Test Plan Overview

The format of the Master Test Plan that we have defined has the same main topics as the recommended approaches, however there are three differences:

- While we have retained the ANSI topics, we have modified what we feel is important for financial institutions when testing software.
- We have added some topics that we feel should be considered.
- We believe that each Master Test Plan has to be customized for each organization, and potentially for each major Application Development Project. Testing Projects are complex enough without requiring the Test Manager to exactly follow some rules that have been established for all industries in all circumstances.

Our Master Test Plan is divided into three major sections:

Summary Test Plan

The development of the Summary Test Plan is covered in the balance of this chapter.

Detailed Test Plans

Detailed Test Plans should be created for each Project Phase. The Summary Test Plan defines the approach and Scope of the overall testing effort while the Detailed Test Plans support the various Test Types in each Test Phase. This process is described in Chapter VII.

Project Plan

The Test Project Plan starts when the Summary Test Plan is written, and is extended as each of the Detailed Test Plans is created. There is some overlap between the Summary Test Plan and the Detailed Test Plans, with the difference being the level of detail that is covered in each Plan.

The Project Plan is a list of all of the Tasks that are associated with the Planning and Execution of the testing effort with start and end dates. This process is described in Chapter VIII.

Summary Test Plan Contents

The Summary Test Plan should contain a high level review of all the information that impacts the testing Project. The following is the authors' approach to the Summary Test Plan:

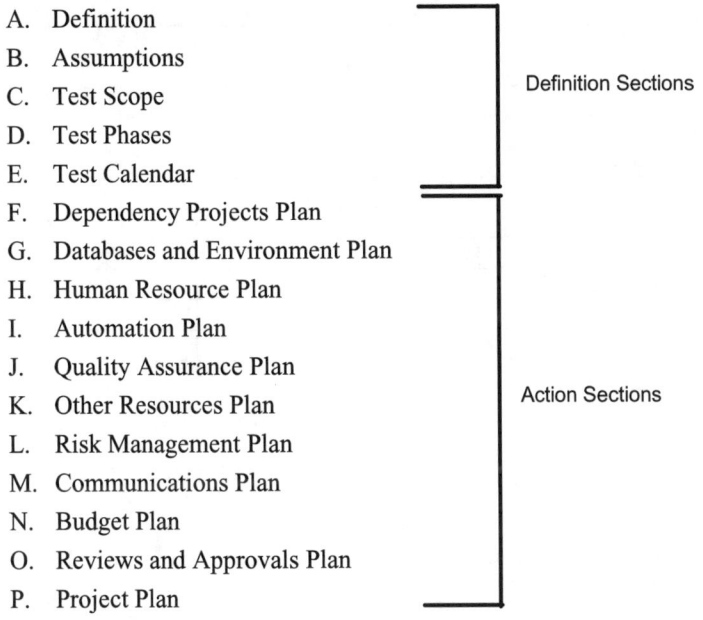

A. Definition
B. Assumptions
C. Test Scope
D. Test Phases
E. Test Calendar
F. Dependency Projects Plan
G. Databases and Environment Plan
H. Human Resource Plan
I. Automation Plan
J. Quality Assurance Plan
K. Other Resources Plan
L. Risk Management Plan
M. Communications Plan
N. Budget Plan
O. Reviews and Approvals Plan
P. Project Plan

Definition Sections (A–E)
Action Sections (F–P)

While the outline of our proposed Summary Test Plan differs from that of the ANSI Standard 829 outline, it contains all of the essential pieces of information for both the internal and external audiences. Additionally, there are seven sections that are in our version of the Master Test Plan that are not in the ANSI Standard 829. They are:

- Assumptions
- Dependency Projects Plan
- Automation Plan
- Quality Assurance Plan
- Other Resources Plan
- Communications Plan
- Budget Plan

The following chart illustrates the differences between an ANSI Standard 829 Plan and the hybrid we recommend.

MASTER TEST PLAN SECTIONS	ANSI STANDARD 829	TSG SUMMARY TEST PLAN	TSG DETAILED TEST PLAN	TSG PROJECT PLAN
Introduction	X	X		
Test Items	X			
Features To Be Tested	X	X	X	
Features Not To Be Tested	X	X		
Approach	X	X		
Item / Pass / Fail Criteria	X		X	
Suspension Criteria and Resumption Requirements	X	X		
Test Deliverables	X			X
Test Tasks	X			X
Environmental Needs	X	X	X	
Human Resource Plan	X	X		
Schedule	X	X	X	X
Risks and Contingencies	X	X		
Reviews and Approvals	X	X		
Assumptions		X		
Dependency Projects		X	X	X
Automation Plan		X	X	X
Quality Assurance Plan		X		
Other Resources		X		
Communications Plan		X		
Budget		X		

Figure 28 – TSG Approach Compared to ANSI 829

Each of these sections will be explained later in this chapter.

Establishing the Summary Test Plan

The following sections describe the type of information that should be covered in the Summary Test Plan.

Definition

Many different people will use the Summary Test Plan as a source Document to write portions of the Detailed Test Plans, and to help the people who are involved in testing understand the goal and Scope of the overall testing effort. There are a number of areas that should be covered in this section:

- What is the business objective of the new Application or Change?
- If a legacy Application is being replaced, how does the new Application differ from the existing Application (Functions, technology, processing Interfaces, etc.?
- How will the End User use this Application or Change once it has been implemented?
- Who is the business owner of the Project and what do they expect?

In addition, the Definition section must identify the Documentation that should be used to write all of the Detailed Test Plans. This Documentation can include

Business and Functional Requirements, User Manuals, Workflows Diagrams and current procedures, and should be easily available to the Test Team managers, preferably in electronic form.

Assumptions

After defining the testing effort, the Summary Test Plan should identify the Assumptions that were made. These Assumptions should be clearly stated and agreed to by all of the affected parties. The Assumptions should cover the following areas:

Testing Roles and Responsibilities

> **Authors' note:** Misunderstood or poorly communicated Assumptions are a leading cause of Project failure.

This section of the Summary Test Plan should identify the Assumptions that were made regarding the testing roles and responsibilities, and it is especially important in the following situations:

- If the Project involves a vendor Application, the Assumptions should clearly state the difference between the testing that is being conducted by the vendor and the testing that is the responsibility of the Test Team.
- If the Application is being written internally, the Assumptions should clearly identify what the Development Team, the Test Team and the End Users will test.
- If Dependency Projects are involved, the Assumption is usually made that the area managing the Dependency Project would be entirely responsible for testing that Application. However, there will be some coordination required where the Applications Interface with each other. As a result, the roles of the Test Teams in the sending and receiving Applications should be clearly defined in the Assumptions section.
- If the Master Test Plan only covers certain Test Types and an Assumption is that the other Test Types will be covered by another area and hopefully another Detailed Test Plan (e.g., if the Unit and System Integration Testing are being handled by the Development Team and that Plan is not included in the Master Test Plan) the Assumptions section should indicate that these Test Types are being covered separately.

Standards

The Test Manager should identify which of the standards that were presented in Chapter V would be employed in the testing effort. Those standards, their Scope and their impact should be defined.

Timeframes

Frequently, there is an end date for the Implementation that is mandated either by the business' need or Regulatory Requirements. When these timeframes exist, they should be outlined in the Assumption section, as this may significantly impact the Scope and Test Types that can be executed, as well as the resources that are needed.

The Master Test Plan could state that it is critical to meet a specific deadline and that certain Test Types could be abbreviated if necessary in order to meet the deadline.

Delivery Requirements

There are a number of situations where assumptions are made about the delivery of the Code. Some of these could be:

Vendor Applications

> **Authors' note:** If the Project has an immovable live date, the Test Manager has to ensure that there is sufficient time to test the Application, and Re-test the inevitable Defects.

Where there is a vendor Application that is either being tested for the first time or Changes that are being Developed for an existing vendor Application, it is usually assumed that some level of testing has been performed by the vendor. This level of testing should be outlined in the Assumption section because this may allow the elimination of some Test Types if they are satisfactorily Planned and Executed by the vendor, such as Unit or System Integration Testing.

Testing by Internal Developers

If the Master Test Plan only covers the testing that is performed by the Test Team and/or End Users, the assumption may be that the Unit or System Integration is covered by a separate Plan created by the Development Team. The Assumption section should clearly state that the testing covered by this Plan and the related Test Calendar assumes that the Unit and/or Systems Testing has been thorough and successfully Executed before the Code is delivered.

Suspension Criteria and Resumption Requirements

The ANSI Standard calls for a firm Suspension Criteria and Resumption Requirements Policy. We feel there should be some Guidelines, but that the Test Manager needs to have the freedom to make a judgment call in suspending and resuming the testing efforts. Because each Project is different, it is very difficult to document what might happen during the Test Phases, the Test Manager should not be constricted by these preliminary guesses.

Responsibilities

The responsibilities of the Test Team should be defined at a high level prior to defining the staffing requirements. However, once the staffing responsibilities and levels have been decided, the Test Manager should have the flexibility to reassign or combine responsibilities. Having a strict definition of the roles and responsibilities can limit the Test Manager's ability to manage in a fluid situation.

Limitations of the Master Test Plan

This section should document any limitations to the Plan itself, including topics that will not be covered in the Plan.

Support

The level of support provided to the testing effort is critical to the success of the overall Project. This support should come from a number of areas, and it is important to define these expectations. A few of the areas providing support are described in the next sections.

Development Team

The Development Team may play a number of important roles in support of the testing effort, not the least of which is the turn-around time for the resolution of Issues. The Test Calendar that is established must make some assumption regarding the turn-around time for Defects and other Issues that are identified during the Testing Phase. It is important to document this expectation.

Business Analysts

It is usually assumed that all of the Design work is completed prior to the beginning of the testing activities; however, the Design is not always frozen before the Test Plan or Test Procedures are written. Depending on the expectations of the Design Phase, it is usually helpful to define the assumptions related to the Design before the Project begins.

End Users

It is important to specify when the testing will rely on support by the End Users. This involvement may come in the form of people who will actively participate in the testing; however, the End Users are almost always needed on an ad hoc basis to resolve Issues that normally surface during the testing process.

In addition to these assumptions, there may be many more that are organizationally specific. Documenting all of the assumptions makes it easier to ensure a common understanding and to prevent misunderstandings when the Project hits its first, inevitable, roadblock.

Test Scope

The Scope section in the Summary Test Plan is the first description of what the testing should cover, and it should provide the answers to the following questions:

- What Functions and Application Components will be tested?
- What Functions and Application Components will not be tested?
- In how much detail will each Function and Component be tested?

A Function is an activity that is performed by an Application to support a business requirement. Some examples of Functions in financial services are:

- Buy or sell a security
- Posting a debit or credit to an account
- Paying a dividend or interest
- Making a loan payment
- Issuing an insurance invoice
- Setting up an account
- Moving money from one account to another

The initial list of Functions should form the basis for a numbering scheme, such as:

FUNCTION	FUNCTIONAL NUMBER
Log-on/off	1.0
Systems Administration	2.0
Buy Security	3.0
Sell Security	4.0
Cash	5.0
Receive	6.0

Figure 29 – Example of a Functional Numbering Scheme

The primary Components of an application that need to be tested in any financial application are:

- User Interfaces
- Reports
- Files
- Application Interfaces

The Functions will be tested through the use of functionally-oriented Business Scenarios, and the Components can be tested through the Business Technical Tests.

Business Functional Tests [24]

Business Scenarios are based upon the Application's Functions that are listed in the Summary Test Plan. The process of writing Business Scenarios can begin as soon as the Functions have been identified, and the Test Scope has been established.

Functionally oriented tests are usually performed in the following Test Types:

- Unit Tests
- Systems Integration Test
- Functional Test
- Business Integration Test
- User Acceptance Test
- Regression Tests

---

[24] Business Functional Scenarios are introduced here and further defined in Chapter VII, and a detailed explanation is included in Chapter IX.

The following Matrix is an example of how specific Functions can be tested within multiple Test Types.

BUSINESS FUNCTIONS USED IN TEST TYPES						
	UNIT & SYSTEMS INTEGRATION	FUNCTIONAL	BUSINESS INTEGRATION	USER ACCEPTANCE	PARALLEL	REGRESSION
Functions						
F-3.0 Buy Security	Limited	Full	Limited	Full	Full	Limited
F-4.0 Sell Security	Limited	Full	Limited	Full	Full	Limited
F-5-0 Cash	Limited	Limited	Limited	Full	Full	Limited

Figure 30 – Sample Matrix for Business Scenarios

The differences between the activities in the Test Types in this Matrix are defined as follows:

Not Included

Not all of the Scenarios will be tested in each Test Phase because:

- It may not be possible to test something until all of the related modules are delivered and ready for testing
- All the required data for the Scenario is not available in a particular Test Phase
- It may not be practical to test because of Environmental restrictions

Limited Testing

Limited Testing of Scenarios can occur for a variety of reasons, such as:

- Unit and Systems Integration Testing may not include all of the Functions
- Some of the required data may not be available to test every data scenario
- Certain test Types, such as in the Parallel Test, may not include a wide variety of different transaction types, so the testing is limited to only those activities that occur in Production during the Parallel Test period

Full Testing

Full Testing is the most comprehensive form of testing and includes all of the possible functionally oriented Test Transactions (or as close to 100% as time and resources will allow) that are available for the respective Test Phase.

After the Business Scenarios have been identified, they can be extended into Test Cases and Test Procedures in the Detailed Test Plans. This is covered in more detail in Chapters VII and IX.

Business Technical Tests [25]

In Business Technical Testing, the Test Team uses combinations of data that are logically possible, but which are not derived directly from the Business and Functional Requirements Documents. Business Technical Tests should cover the Application's Components, and they are most likely to be conducted during one or more of the following Test Types:

- Unit Tests
- Integration Test
- Functional Test
- Business Integration Test

The Scope section in the Summary Test Plan should identify all of the Technical Scenarios that will be tested, while the process of writing the actual Business Technical Tests is discussed in Chapters VII and X.

In most Projects there is a series of Technical Conditions that apply to the standard Application Components. Some of the combinations that are found most often are:

BUSINESS TECHNICAL SCENARIOS AND APPLICATION COMPONENTS				
TECHNICAL CONDITIONS	USER INTERFACES	REPORTS	FILES/DATABASE	APPLICATION INTERFACES
Completeness	X	X	X	X
Stability	X	X	X	X
Navigation	X			
Appearance	X			
Field Edits	X	X		
Defaults	X			
Warning and Error Messages	X	X	X	X
Security	X		X	X
Format/Content		X		X
Report Process		X		
Data Specific			X	X
Record/File Specific			X	

Figure 31 – Business Technical Conditions Matrix

---

[25] Business Technical Scenarios are introduced here and further defined in Chapter VII, and a detailed explanation is included in Chapter X.

The Application Components can be tested in multiple Test Types, in different levels of detail, as shown by the examples in the following chart.

BUSINESS TECHNICAL SCENARIOS CONDUCTED IN TEST TYPES						
	UNIT & SYSTEMS INTEGRATION	FUNCTIONAL	BUSINESS INTEGRATION	USER ACCEPTANCE	PARALLEL	REGRESSION
User Interfaces	Limited	Full	Limited	Full	Full	Limited
Reports	Limited	Full	Limited	Full	Full	Limited
Files	Limited	Full	Limited	Full	Full	Limited
Application Interfaces	Limited	Full	Full	Full	Full	Limited

Figure 32 – Sample Technical Scenarios and Test Types

The differences between the activities in each of the Test Types in this figure are defined as follows:

Not Included

Not every Component can/should be tested in each Test Phase.

Limited Testing

Limited Testing may only include a subset of the elements that pertain to this Component. The intent is to identify a sub-set that will provide a reasonably complete test, using fewer transactions, and that will not fully cover all of the common testing Conditions, which are discussed in Chapter X.

When testing the formats of the Components, it may not be feasible to test all of the possible combinations until the entire Application is integrated. For instance, it may be more practical to test just the Conditions related to the layout and format of the Screen, Report or Interface so that the Business Scenarios can be tested.

Full Testing

Full Testing is the most comprehensive form of testing and includes all of the possible Technical Scenarios (or as close to 100% as time and resources will allow).

This Matrix forms a Guideline that can be used later to define the Detailed Test Plans' Testing Scope. The Detailed Test Plan should include additional Matrices that identify the level of testing that will be applied to each of the functionally oriented or technical areas.

Dependency Projects

The Application being tested will usually Interface with one or more other Applications. Very often, these other Applications will require some Changes in order to work effectively. The Summary Test Plan should list all of the Applications that will impact the testing Project in order to develop a complete Plan.

A Matrix to outline the Testing Scope of the Dependency Projects might look like this:

DEPENDENCY PROJECTS						
DEPENDENCY APPLICATIONS	UNIT & SYSTEMS INTEGRATION	FUNCTIONAL	BUSINESS INTEGRATION	USER ACCEPTANCE	PARALLEL	REGRESSION
FD-1.1 General Ledger			X	X	X	
FD-1.2 Daily Valuations			X	X	X	
FD-1.3 DTC Settlement			X	X	X	
FD-1.4 Corporate Actions			X	X	X	X
FD-1.5 Fee Processing			X	X		
FD-1.6 Report Delivery			X	X	X	

Figure 33 – Sample Matrix for Dependency Projects

This Matrix lists of all of the Dependency Projects as well as the Test Phases in which they will be tested. In the Detailed Test Plan, the specifics of the Interface between the Applications will be defined, along with the testing that will be completed as part of the Project and the Test Team's responsibility for the Dependency Project.

Test Phases

The Summary Test Plan should outline a high-level approach for each of the Test Types that will be performed. In this section, the Test Types that will be Executed should be discussed as well as those Test Types that are not being Executed along with the reasons why.

Each Test Type could be conducted as a dedicated Test Phase, or more than one Test Type could be grouped into a Test Phase and performed simultaneously. For instance, it may make sense organizationally to group the Unit Testing and System Integration test into one phase and refer to it as the Development Testing Phase. This is most commonly done when the Developers perform these Test Types and they are not part of the Testing Team's responsibilities.

The description of each Test Phase should include the following information:

Objective

The Objective section should outline the purpose of each Test Phase. The Objectives will be more specific in the Detailed Test Plans, but the Summary Test Plan should give enough detail to be able to differentiate between the Test Phases and, at some level, justify why each phase must be Executed.

Environment

The Environment for each Test Phase should be summarized in this section. While the need for various Test Regions has been discussed in another section, this portion of the Plan identifies the differences between the Test Phases that may occur in the same region. The discussion of the Environment should include the following:

- What data is being used (Production versus Manufactured)?
- How is the set-up being performed?

- Who is responsible for the set-up of the Environment?
- Who is responsible to maintain and update the Environment?

These points should apply to each of the Test Phases, and the specifics for each point should be discussed in the Detailed Test Plans. For example, the discussion in the Summary Test Plan may define the Test Environment requirements in the following way:

TEST TYPE	REGION REQUIREMENTS				
	DEVELOPMENT	FUNCTIONAL 1	FUNCTIONAL 2	DEPENDANT APPLICATIONS	CONVERSION
Unit Tests	X				
Systems Integration Test	X				
Systems Technical Tests	X				
Functional Test			X	X	X
Business Integration Test				X	X
Business Technical Test			X	X	
Regression Test				X	
User Acceptance Test			X		
Parallel Test				X	
Conversion Test					. X

Figure 34 – Example of Region Requirements

Exit Criteria

Each Test Phase must have its own exit or completion criterion that is based upon the Status of the Code. These criteria must be met prior to moving to the next phase. The role of the Test Manager will be to enforce these criteria, or to make adjustments due to time constraints or other Project concerns. In a perfect world there only would be two exit criteria:

- All of the Test Procedures have been Executed, AND
- There are no remaining Defects of a high operational or quality impact

Realistically, time always becomes a factor in deciding when to follow or bypass the exit criteria. To some degree, the completion of a Test Phase will be dictated by the Project timeline. As a result, it will be necessary to specify in the Summary Test Plan where concessions could be made to allow the testing to move forward to the next phase.

One of the possible concessions could be to allow pieces of the Application to move into the next Test Phase without moving the entire Application. However, even with this concession, the expectations regarding the minimum criteria that must be met before moving into the next phase of testing should be documented and followed.

Authors' note: Regardless of the size of the Project or the number of people available, all of the Functions must be performed.

Test Calendar

In writing the Summary Test Plan, the Test Manager should examine all of the Application's Functions and Conditions that require a specific Test Day Condition. For instance, the purchase of a US Equity occurs on one day and a series of events occur over the next two business days until the trade settles on the third business day. During this period, several events can occur that would require that five contiguous business days be run. A simplified view of this is shown in the following Matrix.

	TRADE DAY	TD + 1	TD + 2	TD + 3	TD + 4
Test Scenario	F-3.1 Buy US Equity			Settle US Equity	
Test Case 3.1.1		Cancel			
Test Case 3.1.2			Correct		
Test Case 3.1.3					Cancel

Figure 35 – Example of Test Cases Requiring Multiple Test Days

Types of Test Days

In general, the Test Calendar consists of three categories of days:

- The Test Days that result from what will be tested and can be represented by either:
    - A processing Logical Day (Day 1, Day 2, etc.)
    - An actual Production Day (October 29, October 30, etc.)[26]
- The actual Calendar Days that identify when the Test Days will be run

The Test Calendar can involve a combination of Test Days and Production Days.

SAMPLE TEST CALENDAR		
CALENDAR DAY	TEST DAY	PRODUCTION DAY
Test Cycle 1		
8/29	Day 1 Data Entry	
8/30	Day 1 Evaluation of Results	
8/31	Day 2 Data Entry	
9/01	Day 2 Evaluation of Results	
Test Cycle 2		
9/05		10/29 Data Entry
9/06		10/29 Evaluation of Results
9/07		10/30 Data Entry
9/08		10/30 Evaluation of Results

Figure 36 - Sample Test Calendar

---

[26] The Production Day could either be a current day in a Parallel Test, or consist of Production Data that was saved from a specific day's work and which is used in various Functional Tests.

Test Conditions Affecting the Test Calendar

There are different categories of Conditions that will dictate which Test Days must be run in the course of the testing. For example, if the Application is one that produces Client Statements on a monthly, quarterly and annual basis on the calendar month-end, the Test

> **Authors' note:** Do not initially Plan to work on weekends. You will almost always need that time to catch up when something goes wrong.

Calendar should include at least three month-ends to test the following Conditions:

PROCESSING MONTH	CLIENT STATEMENT		
	MONTHLY	QUARTERLY	YEAR-END
October	X	January/ April/ July/ October	October year-end cycle
November	X	February/ May/ August/ November	November year-end cycle
December	X	March/ June/ September/ December	December year-end cycle

Figure 37 – Example of Test Conditions

Of course, more months would improve the Client Statement tests, but part of the Planning process is to determine what is absolutely necessary to test the Conditions in the least number of Test Days and not over test.

Let's take the previous example and add the requirement to backdate transactions that will be included in the Client Statements. In this case, the Test Calendar may look like this:

TEST CALENDAR	
TEST CYCLE 3	TEST CONDITIONS
10/31	Monthly Client Statement
	Quarterly Client Statement on a J/A/J/O [27] cycle
	Annual Client Statement for an October Year-End cycle
11/01	Enter back-dated transactions for October
11/02	Re-run October Client Statements
11/30	Quarterly Client Statement on a F/M/A/N cycle
	Annual Client Statement for a November Year-End cycle
12/31	Quarterly Client Statement on a M/J/S/D cycle
	Annual on a December Year-End cycle

Figure 38 – Impact of Backdating on Test Conditions

The Test Calendar section of the Summary Test Plan should outline which Test Days will be used in the testing and in which Test Phases they will be Executed. It may not be necessary (or possible) to test every Test Day in each Test Phase. Using the Test Days from the previous example, the Test Calendar by Test Phase may look like the following figure.

---

[27] January, April, July and October

| TEST CALENDAR |||||||
TEST CYCLE 4	UNIT & SYSTEMS INTEGRATION	FUNCTIONAL	BUSINESS INTEGRATION	USER ACCEPTANCE	PARALLEL	REGRESSION
10/31	X	X	X	X		
11/01		X	X	X		
11/02		X	X	X		
11/30		X	X			
12/31		X	X			

Figure 39 – Basic Test Calendar

Once the Project Manager has determined the Test Days that will be needed and the Test Phases in which they will be Executed, the next step is to assign start and end dates to these Test Days and estimate the time requirements for the tests.

Estimating Time Requirements

There are several Rules of Thumb that can used to estimate the time required for the Test Calendar, including:

- The total amount of time required for the testing is at least half as much time as the Requirements and Development Phases combined.
- One-third of the testing effort is used to develop the tests, one-third to execute the tests, and the remaining time is divided between planning and re-testing.
- If an Application is acquired, one-third of the total cost is the cost to acquire the product, one third to customize it, and the remaining one-third is needed to implement it, including testing.

Participants' Track Record

The track record of the Test Team in an established organization should be reviewed to see:

- What percentage of testing time versus Development time has occurred in past Projects?
- How accurate have the Testers' and Developers' prior estimates been?
- Will the same people who have worked on past projects be working on this one?

When testing a vendor Application, the Project Manager should contact other users of the vendor's Application to get their testing history for this Application in similar situations (i.e., number of Changes, quality of Defect Repair, Defect turnaround time, etc.)

Authors' note: To evaluate the pros and cons of building a new team vs. outsourcing, see Chapter XVI.

Resource Availability

The resources that are available will affect the start and end dates for each Test Phase. If the testing resources are working on other Projects, the dates for this Project may be affected based on the priority and status of the other Projects. Also, if a new Test Team or End-User team has to be established, the lead-time for hiring and Training may also alter the start and end dates.

Developing the Timeline

Once the time period required for the entire testing effort has been estimated, the timeline should be calculated for each Testing Phase and Test Day. In order to develop the start and end dates for the actual test activity, the following formula could be used:

Total Days for the Testing Effort / Number of Test Dates

This formula calculates the number of Calendar Days available for each Test Day, which can be used to establish the overall Test Calendar. In order to complete the Test Calendar, the Project Manager will need to balance the number of days that are needed for testing against the end date of the testing as mandated by the overall Project Plan and then add in the other preparatory and support activities.

It may be necessary to make adjustments throughout the testing effort to be able to meet the end date and still perform a successful test.

In the previous example of the Test Calendar (Figure 39), there are fourteen Test Days that need to be run in order to complete all of the Test Conditions in the Test Phases. Based upon the Test Manager's estimates, the overall Project Plan has allocated 70 Calendar Days for testing, and each Test Day would take five Calendar Days. In this case, the Test Calendar would look as follows:

TEST CALENDAR														
TESTS	WEEK NUMBERS													
	1	2	3	4	5	6	7	8	9	10	11	12	13	14
DEVELOPMENT														
Date: 10/31	X													
FUNCTIONAL														
Date: 10/31		X												
Date: 11/01			X											
Date: 11/02				X										
Date: 11/30					X									
Date: 12/31						X								
BUSINESS INTEGRATION														
Date: 10/31							X							
Date: 11/01								X						
Date: 11/02									X					
Date: 11/30										X				
Date: 12/31											X			
USER ACCEPTANCE														
Date: 10/31												X		
Date: 11/01													X	
Date: 11/02														X

Figure 40 – Sample Test Calendar

Adjusting the Timeline

However, if for a variety of externally imposed reasons, the overall Project Plan only allows for ten weeks of actual testing rather than fourteen, the Test Manager will be challenged to identify ways to reduce the total elapsed time for testing. Some of the techniques that we have used to reduce the elapsed time are shown in the following table:

REDUCING TEST CALENDAR TIMEFRAMES		
TECHNIQUE	**ADVANTAGES**	**DISADVANTAGES**
1. Reduce the number of Calendar Days per testing day	Meets the overall deadline Avoids testing confusion by eliminating the need for overlapping days	May have to Change the Scope to accommodate a shorter cycle If Re-testing is required, there is much less flexibility May require multiple teams or shifts May pressure the manager to 'roll' the Test Day before the criteria have been met
2. Allow the Test Days to overlap	Avoids reducing Scope	Becomes more complex Requires more resources for Test Execution, administration and Environmental Control May require more Test Regions
3. Increase resources	Maintains the existing Plan and Scope	Will increase the cost of the Project May require additional Training for the new people Will require additional support resources (space, PCs, etc.) May increase the complexity of the Project
4. Change the Scope	Simplest management Task	May increase the Risk when the Application moves to Production

Figure 41 – Reducing the Test Calendar Timeframe

Each of these four techniques is discussed in the following sections:

Technique 1 – Reduce the Calendar Days per Test Day

In order to reduce the Calendar Days that are required to complete the Test Day, the following options should be explored:

- Expand the role of automation in the testing
- Add resources for input or validation Tasks (See Technique 3)
- Add resources and test in shifts (See Technique 3)
- Reduce the Scope (See Technique 4)
- Combine Test Cases

Most of these options will require additional funding so the Test Manager will need to ascertain if this is a viable option.

Technique 2 – Overlap Test Days

The following example shows the impact of overlapping Test Days. In this example, if there are sufficient Test Regions to support concurrent testing activities, the Business Integration Test could start a week earlier (before the Functional Test has been completed), and the User Acceptance Test could begin before the Business Integration Test has been completed.

In this case the Test Calendar would look as follows:

TESTS	\multicolumn{12}{c|}{TEST CALENDAR – WEEK NUMBERS}											
	1	2	3	4	5	6	7	8	9	10	11	12
SYSTEMS INTEGRATION												
Date: 10/31	X											
FUNCTIONAL												
Date: 10/31		X										
Date: 11/01			X									
Date: 11/02				X								
Date: 11/30						X						
Date: 12/31							X					
BUSINESS INTEGRATION												
Date: 10/31						X						
Date: 11/01							X					
Date: 11/02								X				
Date: 11/30									X			
Date: 12/31										X		
USER ACCEPTANCE												
Date: 10/31										X		
Date: 11/01											X	
Date: 11/02												X

Figure 42 – Sample Test Calendar (Overlapping Phases)

If the same resources are required for the overlapped tests, additional people may be required.

Technique 3 – Increase Resources

Additional resources could be obtained from a variety of sources, including:

- Additional End Users or Developers
- Additional Business Analysts from another project
- Consultants
- Outsource portions of the test [28]

Technique 4 – Change the Scope

With the concurrence of the Project Management Team, the Test Manager can reduce either the number of Functions that are tested or the number of Test

---

[28] Chapter XVI discussed the pros and cons of Test Outsourcing.

Days that are included in each Test Cycle. This is a major Change to the Project and increases the Risk that some serious Defect will still exist after the Application is moved into Production.

As shown, there are many ways to adjust the Test Calendar to fit within the overall Project Plan. It is the Test Manager's responsibility to establish a Test Calendar that will allow for a complete test while meeting the constraints of the Implementation date.

> **Authors' note**: Scope modifications should only be made with the agreement of the Project Manager and senior management – and must be documented.

Everyone involved with the Project should be aware that Changes to the Test Calendar are inevitable. Also, regardless of how complete the Test Calendar is when the Project is started there will be events that occur that will cause the Test Manager to re-think the process. In most cases this will require compromises in one or more areas, such as:

- Management expectations
- Test Scope
- Test Calendar and/or delivery date
- Incremental deliveries

Databases and Environment Plan

Each of the Test Phases may have its own requirements for the Test Environment, which could include:

- Test Regions
- Interfaces with other Applications
- Databases
- Parallel Environment

Each of these requirements is discussed in the following sections:

Test Regions

It is important to review the Test Environment requirements in the Project Management Meetings so that the impact of any decisions that restrict the number or size of the Test Regions can be understood by all of the participants and interested parties.

In Medium and Large Projects, it is important to have more than one Test Region or Environment. Often there is at least one Test Region for the Development Team and at least one for the Test Team. The Development region allows for the Unit Testing and Regression Testing of Defect repairs without impacting the Test Region. In some of the larger Testing Projects it is often necessary to have multiple Test Regions so that concurrent testing activities can take place, such as preparation for Parallel Testing in one region while Business Integration Testing is going on in another.

The Summary Test Plan should list all of the Test Regions that will be needed, how they are being used and in what timeframe. A Gantt chart can very easily show this as the following example illustrates:

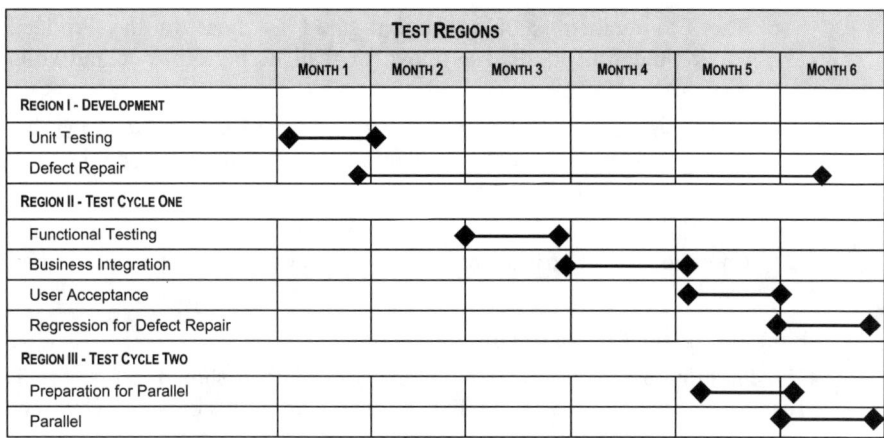

Figure 43 – Sample Use of Test Regions

Interfaces with Other Applications

When testing an Interface, the Test Manager may need one region for the tested Application and another for the interfacing Application. If so, it is important to list all of the Applications that will need a separate Test Region in order to effectively test the Interface. The Summary Test Plan should also document any constraints that are placed on the testing because of these Interfaces or a lack of available Test Regions.

For example, it may be that the interfacing Application can only be accessed on certain dates. If that is the case, it is important to list all of the limitations and identify the impact on testing.

> **Authors' note:** If you are testing an Interface to an Application that requires its own region, and you want to run some Functional tests concurrently with the Business Integration Test, you will need at least four Test Regions.

Databases

In many Projects, there is a need for more than one Database. Some of the reasons for this are:

- Different versions of the Application may be in test at the same time, and they may require different sets of data
- Tests in different Test Regions may require independent Databases
- If Dependent Applications are in different regions for reasons of their own, different Databases may be required

Each Database must be managed so that it is synchronized with the other Databases, or that any differences are known and accepted.

Parallel Environment

When a Parallel Test is being Executed, the Environment for this testing is often different from the rest of the Test Regions for a number of reasons, including:

- The Environment should be configured as close to the Production Environment as possible to allow for realistic performance and volume results.
- The Environment should have significant technical support so that it can realistically cycle through a Business Day in a real-time mode (i.e., to run a Business Day in a Calendar Day).
- The Environment will need all connections to interfacing Applications (in either a Test or Production Region) that the Production Environment will have.

Back-ups of the Environments

The Environment section of the Summary Test Plan should define the Test Teams' expectation of how the Test Regions will be backed-up. This should include:

- Schedule for back-ups (based on the Test Calendar or the Test Day)
- Types of back-ups (full or partial)
- Retention of back-ups
- Restoration protocol and timeframes

These items may be different for each of the Test Regions, and they should be documented in the Summary Test Plan so that the expectations and responsibilities are clear.

As with each of the sections of the Summary Test Plan, the Environment portion should contain all of the information that is relevant to the Project. It is important to include this information so that the Testers, the Development Team and other support resources understand that the Test Plan was established with specific expectations on the availability of the Test Regions, and any Change could alter the Master Plan and the Test Calendar.

Data Sources

Each of the Test Regions may have different sources for the data that is used to set-up Accounts, Static Files, etc., and for the Test Transactions. In most cases the data comes from one of three places:

- Automated Conversion of Existing Data
- Manual Set-up
- Copy of a Production Environment

It is important to understand the source of the set-up data because this decision can seriously impact the Test Calendar, as well as the resources that are required to perform the set-up Tasks. Some of the set-up data may be FX rates, BIC codes, FIN codes, securities descriptions, accounts, positions, account history, etc. Depending on the data requirements for the Testing Phases, preparing the Test Region(s) can be quite time-consuming and a significant use of Test Team and IT resources. This section of the Master Test Plan should speak to the source of the set-up data.

In addition, the transaction data that will be used for testing may be obtained from the Production Environment in the case where the Application is replacing a legacy Application, or the Test Team could manufacturer the data.

Chapter VII, which describes the Detailed Test Plan, will address this in greater detail for each phase of testing because it is very often different for each phase. However, the Summary Test Plan should indicate whether the transaction test data in each of the Test Regions would be Production Data or Manufactured Data for the various Functions.

TEST REGION	FUNCTION	MANUFACTURED DATA	PRODUCTION DATA
Test Region I	Buy Security	X	
	Sell Security	X	
Test Region II			
	Interface to Settlement Application		X

Figure 44 – Example of Data Usage by Test Region

There are different benefits and issues associated with the use of Manufactured and Production Data.

	MANUFACTURED DATA	PRODUCTION DATA
ISSUES	Takes time to assemble accurate combinations of data	Production Data may be difficult to access and may require some mapping of Data Elements before it is useful to the new Application
	May be difficult to manufacture certain categories of Test Transactions, such as SWIFT Messages	All of the data elements required to initiate a transaction may not be available in the product files
BENEFITS	Manufactured Data may not cover every Condition	Covers Conditions that have been experienced in Production that might not be considered when the data is manufactured
	Possible to create arrays of tests with defined differences so that Testers can move from simple to complex tasks	Using Production Data ensures that the Messages are already in the correct format

Figure 45 – Comparison of Sources of Transaction Data

Human Resource Plan

Staffing

The Test Team has the following responsibilities:

- Create Business Scenarios, Test Cases, Test Procedures and Expected Results
- Execute the tests according to Plan
- Evaluate the Test Results
- Log Defects and Design Issues
- Prepare data for automated test tools
- Prepare the Test Environment
- Support End Users during their Testing Phases
- Review User Manuals for accuracy
- Assist with Defect resolution

> **Authors' note:** Planning that Testers will 'make up time' later is just wishful thinking. Since you can only Plan for what you know, Testing Projects are much more likely to need more time than Planned, rather than less.

While meeting these responsibilities, the Test Team might perform many different roles. Chapter II discusses building the Test Team and defines the roles that are played within the Team. After the roles have been defined, the Project Manager must estimate the number of people required to perform these activities.

If a Test Team is already in place and the resources are fixed, it may be necessary to adjust the Test Calendar to reflect how the Test Days can be Executed given the number of Testers available. However, if the Test Team is not established or if Testers need to be added to an established organization, the Test Manager will have to estimate the number of resources required, and then obtain the people either internally or externally. [29]

There are several ways to create a staffing Plan, each of which can be used when some of the variables are fixed:

Work Within the Available Budget

Dividing the available Budget by the average cost of an employee or a consultant will yield the number of people who can be applied to the Project. This tops-down approach yields the timeframe that will be required.

$$PR = AB / AC$$

Staffing Based upon Timeframes

When the timeframe is fixed, the method to determine the number of staff required is based upon the total effort that is required, divided by the effort per Task, while considering the total amount of time that is available.

$$PR = (ER / TR) / TA$$

Budgeting Based Upon Requirements

The Test Manager can also determine the Tasks required to complete the overall effort, assign staff and to calculate the amount of time that will be required to complete the Task. The Budget that is required can be calculated from this.

$$\text{Method 1} \quad RB = PR \times AC \times TA$$
$$\text{Method 2} \quad RB = (ER/TR) \times AC$$

The most difficult situation is when the time allocated for testing in the Overall Project Plan is fixed, and there are not enough people or Budget to perform all of the required tests in the available time. In this case, the Test Manager must adjust the Test Calendar as discussed earlier in this chapter.

Once the level of resources is determined, the Test Manager should then allocate the resources to each of the Test Types and Test Phases. The Summary Test Plan should identify the resources at a generic enough level to allow the Test Manager the freedom to move resources between roles or phases as needed. The following is an example of how the Resource Allocation may look:

---

[29] Internal resources are better in that they probably know the business and are already budgeted. The downside is that they can be easily pulled away when some crisis occurs in another area. The positive aspect of using external Testers is that they will focus only on the testing effort, and are usually more willing to "roll up their sleeves to get the job done."

RESOURCE ALLOCATION													
TESTING RESOURCES		WEEK NUMBERS											
		1	2	3	4	5	6	7	8	9	10	11	12
Management														
Test Manager		1	1	1	1	1	1	1	1	1	1	1	1
Administration			1	1	1	1	1	1	1	1	1	1	1
	Sub-total	1	2	2	2	2	2	2	2	2	2	2	2
Development														
Unit Testers		4	2	2	1	1	1	1					
	Sub-total	4	2	2	1	1	1	1					
Functional														
Testers		3	4	4	3	2	1	1					
	Sub-total	3	4	4	3	2	1	1					
Business Integration														
Testers				2	3	4	4	4	3	1	1		
	Sub-total			2	3	4	4	4	3	1	1		
User Acceptance														
Testers									2	3	4	4	4
	Sub-total								2	3	4	4	4
Total Resources		8	8	8	8	8	8	8	8	8	7	7	6

Figure 46 – Sample Resource Allocation

In Planning how many people will be needed, the Test Manager should consider several factors, including:

- The Testers must prepare the Environment and the Test Procedures prior to the actual Test Execution in each Test Phase.
- Enough lead-time should be allowed for new people to be hired and trained prior to the testing activities.
- The resources should remain with the Test Phase after the original Execution is complete in order to perform Regression Testing for any Defects that are fixed.
- People are also needed to perform any ad hoc testing that the Testers that was not originally specified.
- Resources need to be allocated and not have drastic resource peaks and valleys, since this is not practical in any organization or Project.

Training

The Project Manager must consider the Training that is required for the End Users who will ultimately use the new Application or the Change, and also should consider the Training that is necessary for the Test Team. The Testers must understand how to use the Application in order to exercise it. In most cases this means that the Testers will have to be Trained in how to use the Screens and navigate through the Application.

In addition:

- If the Project is going to use a new automated testing tool, the appropriate people may have to be Trained in the use of the tool(s).

- If End Users are assigned to the Testing effort, they might have to be Trained in the Test Process.

Automation Plan

The level of automation is an important factor to consider when identifying the timeline, resources and Testing Scope. Details regarding what tools have been selected, how they will be used and what resources are needed to run and support the Implementation of the automation tools must be defined. The selection and use of the tools are discussed in Chapters XIII and XV.

Quality Assurance Plan

The Test Manager should determine how he/she will ensure the overall quality of the testing effort. This approach may have to be coordinated with the Project Manager's methodology, or the Test Manager may have to establish it just for the Test Plan.

If the Test Manager is establishing a Quality Program, it should be based upon the background presented in Chapter I, and the Guidelines that are reviewed in Chapter XI. The approach should be defined in the Summary Test Plan so that everyone understands what needs to be done.

Other Resources Plan

In addition to the Human Resources, the Test Manager needs to identify other categories of resources that will be required, including:

- Hardware
- Software
- Premises
- Phones, faxes and other communications equipment
- Support

Risk Management Plan

The process of Risk Management is discussed in Chapter XII - Managing Test Administration Processes. The approach that is selected and the procedures should be defined in the Summary Test Plan.

Communications Plan

Based upon the Project Manager's requirements and the Test Manager's approach to managing, The Test Manager has to:

- Agree with the Project Manager on the content, format and frequency of the Reports that are presented to the Project Manager
- Define the content, format and frequency of the Reports that will be used within the testing process
- Identify how information will be collected
- Define the Report production process
- Assign people to do the work
- Define the purpose, frequency and participants for all of the required testing meetings

- Define who will document the results of the meetings and how the results will be disseminated
- Define a central Document repository, if required

All of this should be covered in the Summary Test Plan

Budget Plan

The Test Manager must develop a comprehensive Budget for the testing process and should work with the Project Manager to define the content, format and frequency of the Budget Reports.

Reviews and Approvals Plan

Once the Summary Test Plan has been completed, it should be reviewed by the Project Management Team, Test Team, Development Team, and selected End Users. A walkthrough of the Plan with all of the interested parties can be very useful in order to identify any Issues and jointly resolve them.

After all of the parties have agreed to the Plan, a formal sign-off should occur. The Summary Test Plan should be jointly approved because this is the basis for all the future Planning and Testing.

> **Authors' note:** Most of the testing effort is a collaborative effort. Sign-offs should be obtained to as the events occur.

Project Plan

The last key element of the high level Planning is establishing a summary Task list, which will become the basis for the Project Plan. The Project Plan will evolve as the Detailed Test Plans and the other test Documents are completed; however, the first draft can be started as soon as the Test Manager is in place.

At this point, the Project Plan should have the following information for each task:

- Task Name
- Projected Start Date
- Projected End Date
- Actual Start Date
- Actual End Date
- Responsible Party
- Resources Required
- Predecessor Task

This is the minimum amount of information that is needed in the Project Plan, but depending on how it will be used by the Test Manager, other pieces of information may be added, such as dependencies, percent completed or duration for each Task.

The following is an example of how the Project Plan could look during the Summary Test Planning stage of the process:

	TEST PROJECT PLAN							
	Task Name	Planned Start	Planned Finish	Actual Start	Actual Finish	Predecessor Task	Assigned To
1	**Write Summary Test Plan**	7/01	7/31				
2	Definition	7.01	7/10				Test Mgr
3	Assumptions	7.05	7/31			2	Test Mgr
4	Scope	7/06	7/15				Test Mgr
5	Test Phases	7/08	7/20			2, 4	Test Mgr
6	Test Calendar	7/10	7/22			2	Test Mgr
7	Dependency Projects	7/10	7/23			2	Test Mgr
8	Environmental Requirements	7/15	7/25				Test Mgr
9	Human Resource Plan	7/25	7/31			4, 7	Test Mgr
10	Automation Plan	7/15	7/31				Dev Mgr
11	Quality Assurance Plan	7/22	7/31				Test Mgr
12	Other Resource Requirements	7/22	7/31				Test Mgr

Figure 47 – Sample Test Project Plan

The Project Plan will start with the Planning Tasks that are required for the Summary Test Plan as shown in this example, and expand as more Tasks are identified as the Detailed Test Plans are written. The Project Plan is a very important tool for the Test Manager. It allows him/her to manage the testing process and to identify the Status of the Tasks. Chapter VIII will go into more detail as to how a Test Project Plan can be used as a management tool.

Summary

There are three objectives of the Master Test Plan:
- Document the Test Scope
- Provide Management and Organizational Tools to the Test Team
- Provide Information to Interested Parties

There are three parts to the Master Test Plan:
- Summary Test Plan
- Detailed Test Plans
- Project Plan

VII. DETAILED TEST PLANS

Chapter VI introduced the three main parts of the Master Test Plan and covered the first topic in depth:

- The Summary Test Plan
- The Detailed Test Plans
- The Project Plan

This chapter discusses the next step in the testing process, building the Detailed Test Plans for each Test Phase, and the next chapter identifies what is needed in the Project Plan.

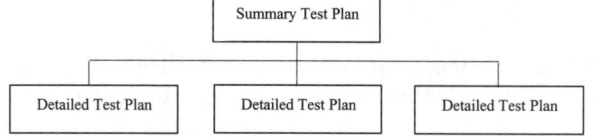

Figure 48 – Relationship of the Summary Test Plan to the Detailed Test Plans

As mentioned in Chapter VI, a Test Phase usually corresponds to an individual Test Type, but a Test Phase can be:

- An individual Test Type
- A portion of a Test Type
- A combination of Test Types

Each Detailed Test Plan should consist of the following sections:

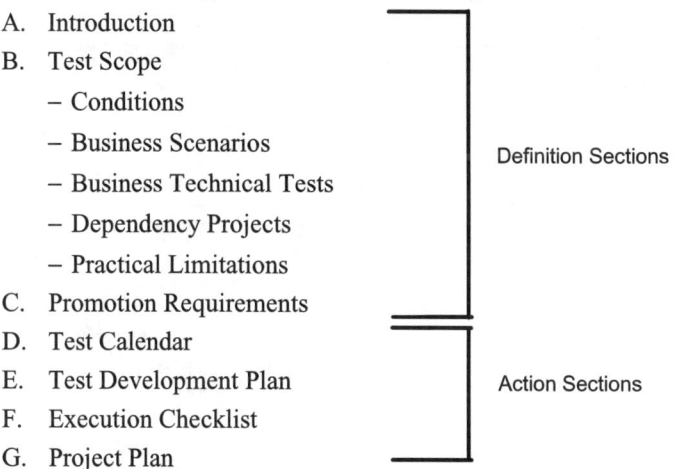

The overall approach to testing is defined in the Summary Test Plan and is expanded as a Detailed Test Plan is established for each Testing Phase. The Detailed Test Plan for a phase will drive the preparation and Execution of the tests that are conducted in that phase.

Authors' note: Rather than repeat the process for each Test Phase, the examples in this chapter will only refer to the Business Functional Test since it requires the most coverage and depth of testing.

However, the Development of any Test Plan is not always linear. In many cases, the definition of some details may cause the Summary Test Plan to be re-examined and possibly modified. And, any modifications to the Summary Test Plan can affect one or more of the Detailed Test Plans.

Establishing Detailed Test Plans

The following sections should be completed for each Detailed Test Plan that is required.

Introduction

The Introduction should provide some background regarding:

- What is to be tested in this Test Phase?
- How does this Detailed Test Plan fit into the overall Master Test Plan?
- What are the objectives of this Detailed Test Plan?

Scope

The major Functions that need to be tested are listed in the Summary Test Plan. This list is then further defined at increasing levels of detail within each Detailed Test Plan.

One technique to assist in establishing the outline is the use of a series of Matrices. This chapter includes examples of how the Business Functional Tests and Business Technical Tests can be expanded to establish the Test Cases that will be needed.

Conditions [30]

The Test Manager should identify the Conditions that will be used to extend the Business and Technical Scenarios into the Test Cases. These Conditions are referenced at several points throughout the book, and can include:

TECHNICAL CONDITIONS [31]	BUSINESS CONDITIONS [32]

Data Conditions
• Positive and Negative Conditions
• Alpha-numeric
• Boundary (Min/Max)
• Zeros
• Negative numbers
• Spaces
• Blank Fields
• Date Formats
• Mandatory/Optional
• Special Characters
	Functional Conditions
	• Transaction Types
	• Instruments
	• Settlement Alternatives
	• Processing Days
	   – End of Day
	   – Weekend
	   – Month End
	   – Quarter End
	   – Year End
	   – Multiple Days
	Volume Conditions
	• High volumes of users performing concurrent activities
	• High volumes of Messages in and out of the Applications
	• High volumes of on-line transactions
	   – In a specific period of time
	   – Over a sustained period of time
	   – Just prior to a cut-off time
	   – Within a narrow processing window
	• Large data bases with complex queries in narrow timeframes
	Processing Conditions
	• All relevant organizations included
	• Day's work be performed in a day
	• Conduct error processing (reject re-entry) in limited timeframes

---

[30] The Conditions listed on these sample Matrices are not meant to be all-inclusive or generic, and the Test Manager will need to determine the unique Conditions that apply to the Project.

[31] Technical Conditions are discussed in depth in Chapter X – Business Technical Testing Process.

[32] Business Conditions are discussed in Chapter IX – Business Functional Testing Process.

	Logical Conditions • Add • Validate • Reject • Modify • Cancel • Reverse • Formulas • Rule-Sets • Code-Values • Backdate • Late Transaction • Return from Counter-parties
User Interfaces • Completeness • Stability • Navigation • Appearance • Field Edits • Defaults • Warning and Error Messages • Security	User Interfaces • Warning and Error Messages • Format • Content • Languages
Reports • Completeness • Stability • Field Edits • Warning and Error Messages • Format • Report Process	Reports • Warning and Error Messages • Format • Content • Report Generation Frequency
Files and Databases • Completeness • Stability • Data Specific • Record/File Specific • Warning and Error Messages • Security	Files and Databases • Warning and Error Messages • Format • Content
Application Interfaces • Completeness • Stability • Format • Data Specific • Warning and Error Messages • Security	Application Interfaces • Warning and Error Messages • Format • Content • Various Message Standards

Figure 49 - Categories of Testing Conditions

The actual processes of creating Business Functional Tests and Business Technical Tests are covered in Chapters IX and X.

Dependency Projects

The Project Test Plans for each Dependency Project should be available to the Test Manager and use the same Project Management approach so that the various Dependency Project Test Plans can be integrated with the overall Test Project Plan.

Practical Limitations

When establishing the Test Scope, the reality of any testing Project is that it is impractical and almost impossible to test everything. It is important to make the audience of the Test Plan and the Testers understand this. There are many reasons why everything cannot be tested, including:

> **Authors' note:** It is impractical to test everything. The successful Test Manager will test enough to ensure a quality Application without over testing.

Business Conditions

Even with a very thorough Detailed Test Plan, The Test Team may not cover every possible business Condition, nor is this an effective investment of time or effort. However, a systematic approach to Functional Testing will significantly reduce the level of Risk associated with implementing a new Application.

Volume

In order to test every aspect of an Application, the Test Team would have to test every combination of the business and technical Conditions for every Transaction Type. For a Medium to Large Project, this would take such a tremendous expenditure of time and money that the Project could not be cost justified.

Human Factors

For GUI Applications, it is not possible to predict every combination or sequence of keystrokes that any user could use on any Screen, nor could the Test Team write a test for every combination.

Given that the Test Team cannot test every combination of business and technical Conditions and Screen inputs, it is important to define tests that will provide the most impact.[33] To define these tests, the Test Team must understand the Application and the business. Once this is understood, the testing can focus on the Transaction Types most likely to be used, and the combinations of Conditions and Business Scenarios that occur most often in Production, or have proven to be problems in the legacy Application that is being replaced.

> **Authors' note:** The Test Manager should not make the decision to skip certain tests without consulting other members of the Project Management Team.

Transaction Coverage

The Functions defined in the Functional Requirement Documents are usually classified as Transaction Types in an Application. When an Application replaces an existing one, the Test Manager can analyze the number of transactions that historically have been processed for each Transaction Type.

Examining Transaction Type usage can help the Test Manager focus his/her efforts. The Test Manager should define a set of Matrices that identify the Transaction Types that are to be covered, realizing that it will be impractical for the testing, either in any particular phase or in its entirety, will cover every Transaction Type.

> **Authors' note:** One very important Task in building these Matrices is to understand the current business or projected business that will use the product.

Very often, 90% of the Application's activity uses 10% of the Transaction Types. For instance, in a trading Application, the two most

---

[33] If the Project Management Team accepts that not every Condition can be tested, it is important to have a process in place after the Application is moved into Production that can Track and Report Defects, and Re-test efficiently.

common Transaction Types are Buy and Sell, although hundreds of Functions exist. The Project Manager should determine the Transaction Types most frequently used during processing, and those should be tested the most.

The following is a sample Matrix that shows how the Transaction Types could have been used in a trading or accounting Application:

TRANSACTION TYPE MATRIX			
FUNCTION	COVERED IN TEST	PRODUCTION USAGE FOR 12 MONTH PERIOD	CUMULATIVE
F-3.0 Buy Security	X	22.50%	22.50%
F-4.0 Sell Security	X	22.20%	44.70%
F-5.0 Cash	X	21.24%	65.94
F-10.0 Dividend	X	5.08%	71.02
F-11.0 Interest	X	4.94%	75.96

Figure 50 – Sample Transaction Coverage Analysis

Depending on time and resource constraints, the Test Manager might decide to not test any Transaction Type that occurs in Production less than 1% of the time, or might use some other criteria. Having this type of analysis allows the Project Manager to make an informed decision that can then be presented to any outside organization that may review the testing, such as Auditing.

Beyond these criteria, the Test Manager may use some of the Transaction Types along with the Business Technical Tests to improve the Test Cases, but the key point for the Detailed Test Plan is to focus on those elements that will most impact the process.

Documenting the limitations of the Scope will allow everyone related to the testing effort to review and validate the decisions that are made by the Test Team regarding what will and will not be covered in the Detailed Test Plans. Of course, the Scope for each Test Phase could be different, but if the same Matrix format is used for each phase, it will be very clear how each Test Phase differs.

Test Calendar

The Test Calendar that was started in the Summary Test Plan is extended in each Detailed Test Plan. The Test Calendar for each Test Phase may be the same with regard to which Test Days are being executed, but the activities that are required in each Test Day may be very different. Therefore, the time allotted for each Test Day can be different for each phase.

Many of the Project's primary target dates were established in the Master Test Plan. In the Detailed Test Plan, additional dates are defined for each Test Phase and the Test Days that are planned. Based upon the available resources, target dates are also established for the support activities that are needed to execute a Test Day, such as data entry, batch processing, and reporting.

Test Development

There are two Test Types that are required in this Test Phase.

Business Functional Tests

The preparation of the Test Cases that support Functional Testing is covered in Chapter IX – Business Functional Testing Process.

Business Technical Tests

The Detailed Test Plans will also identify which Business Technical Test Cases are required to test the appropriate common business Conditions. The preparation of tests for these common Conditions is discussed in Chapter X.

The challenge for the Test Team is to examine the information contained in the Scope section of the Detailed Test Plan and create tests that will cover many different Test Conditions within one Test Procedure, so that the Execution time is minimized.

The Test Cases and Test Procedures should be written as a part of the Detailed Test Plan during the Test Phase that includes the Functional Test, and reviewed for the other Test Phases. The Test Manager can use these tests to establish an initial Test Calendar and assign the tests to particular Test Days so that there is an even distribution of tests and the completion dates can be achieved.

As the Test Procedures are written, additional Test Cases will probably be identified and this might affect the Test Calendar and Project Plan.

After the Test Cases are defined, they can be mapped against the Test Calendar to create the Execution Checklist. This checklist becomes a very important management tool. The Test Calendar in the Summary Test Plan can look like this for the Functional Test:

Test Phase	SUMMARY TEST PLAN - TEST CALENDAR											
	Week Numbers											
	1	2	3	4	5	6	7	8	9	10	11	12
Functional												
Date: 10/31	X											
Date: 11/01		X										
Date: 11/02			X									
Date: 11/30					X							
Date: 12/31						X						

Figure 51- Sample Test Calendar (Master Plan - Functional Test)

The Test Calendar for the Functional Test's Detailed Test Plan for the first two Test Days may look like this:

DETAILED TEST PLAN – TEST CALENDAR										
Functional	9/09	9/10	9/11	9/12	9/13	9/16	9/17	9/18	9/19	9/20
Date: 10/31										
F-2.1 Trade Entry	X									
F-9.0 Settlement	X									
Run Batch Process		X								
Reports Production			X							
Test Verification	X	X	X	X	X					
Date: 11/1										
F-2.1 Trade Entry						X				
F-9.0 Settlement						X				
Run Batch Process							X			
Report Production								X		
Test Verification						X	X	X	X	X

Figure 52 - Sample Test Calendar (Detail Plan - Functional Test)

Execution Checklist

As the Test Cases and Test Procedures are written and mapped to the Test Calendar, they can be added to the Execution Checklist for the Test Manager or Section Leaders.

One additional piece of information that should be included in the Execution Checklist is the name of a responsible party. This will allow the Test Manager or Team Leader to know who to go to for the Status on a particular Task. The Execution Checklist for the first day of testing in the Functional Phase may look like the following example:

EXECUTION CHECKLIST			
SYSTEM DATE	ACTUAL DATE	TEST PROCEDURE / TASK	TESTER
10/31	9/09	F-3.1.1.1 Buy Security - Equity	Terri Tester
		F-3.1.1.3 Buy Security - Equity	Terri Tester
		F-9.1.1.1 Settlement	Tom Tester
		F-9.1.1.3 Settlement	Tom Tester
		Notify Technology of completion of input for the day	Mike Manager
		Process Pricing Files	IT Department
		Run Trade Batches	IT Department

Figure 53 – Sample Execution Checklist

The Execution Checklist is a living Document and will evolve constantly throughout the life of the testing Project. The Test Manager can use this

Document to determine if Tasks start to miss their target dates and if adjustments to resources and timing are needed.

Promotion Requirements

In a perfect world, all of the testing would be completed prior to promoting the Code to the next Test Phase; however, this is rarely the case. For that reason, it is important to document in the Detailed Test Plans the approach that will be used to promote the Code. The promotion Guidelines could be any combination of the following examples, or any that the firm has established.

- Code cannot be promoted if:
  - There are any Critical Issues outstanding before the Code is promoted
  - If more than 2% of the test Cases have not been completed
- Code can be promoted if:
  - All of the Critical Issues are being worked on and the Development Team has a delivery schedule in place
  - There is a work-around available for all High And Critical Issues

Whatever is outlined in this section of the Document, the Test Manager should have the flexibility to make modifications to the Guidelines in order to stay within the timeframes of the Project.

Project Plan

A Test Project Plan is expanded as each Detailed Test Plan is completed. The following example shows how the Project Plan that was started in the Summary Test Plan can be extended at this point.

	TEST PROJECT PLAN							
	Task Name	Planned Start	Planned Finish	Actual Start	Actual Finish	Predecessor Task	Assigned To
1	**Write Summary Test Plan**	7/01	7/31				
2	Definition	7.01	7/10				Test Mgr
3	Assumptions	7.05	7/31			2	Test Mgr
12	Other Resource Requirements	7/22	7/31				Test Mgr
13	**Write Detailed Test Plans**	8/01	9/13			3	
14	Unit Test Plan	8/01	8/07				Dev Mgr
15	Systems Integration Test Plan	8/08	8/11				Dev Mgr
16	Systems Technical Test Plan	8/10	8/15				Dev Mgr
17	Functional Test Plan	8/15	9/4				Test Mgr
18	Introduction	8/15	8/18				Test Mgr
19	Test Scope	8/18	8/20				Test Mgr
20	Promotion Requirements	8/20	8/22				Test Mgr
21	Test Calendar	8/22	8/27				Test Mgr
22	Test Development	8/27	9/01				Test Mgr
23	Define Test Cases	8/28	9/01				Test Mgr
24	Execution Checklist	9/01	9/04				Test Mgr
25	Project Plan	9/01	9/12			21,24	Test Mgr
26	Business Integration Test Plan	8/22	9/11				Test Mgr

Figure 54 - Sample Test Project Plan

The Test Project Plan will be further expanded in Chapter VIII.

Summary

The Detailed Test Plans will provide the information that is needed to perform each Test Phase. The format of the Plan should be the same for each Test Phase, but the details will differ.

Each Detailed Test Plan should contain the following sections:
- Introduction
- Test Scope
- Promotion Requirements
- Test Calendar
- Test Development Plan
- Execution Checklist
- Project Plan

Each Detailed Test Plan will also provide the information the Test Team needs in order to create the rest of the test material, such as the Test Cases and Test Procedures. These will be covered in Chapter IX.

VIII. TEST PROJECT PLAN

The third major part of the Master Test Plan is the Test Project Plan, and it is a very important tool in Planning and managing the testing effort. The Test Project Plan was initiated in the Summary Test Plan, and evolves as a result of:

- The creation of the Detailed Testing Plans
- Modifications to the Test Calendar
- Addition, completion or modifications of the Tasks

The Test Project Plan is built initially upon the information provided in the Summary Test Plan and the Detailed Test Plans, and is extended as the Test Calendar evolves.

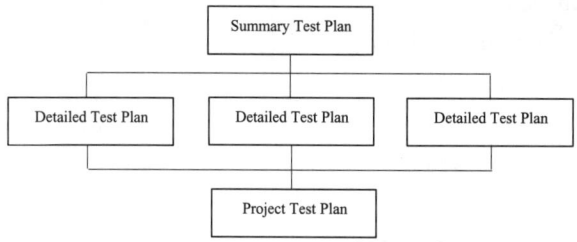

Figure 55 – Example of the Relationship Between the Detailed Test Plans and the Test Project Plan

The Test Project Plan is different from the Test Calendar in that it reflects all of the Tasks that are associated with the testing process, not just the Execution of the tests. It includes all of the Tasks for Planning, Preparation and Execution as well as internal and external dependencies. It is here that the Test Manager can see a complete picture of the testing effort from start to finish and how the resources are matched against all of these Tasks.

> "Effective leadership is putting first things first. Effective management is discipline - carrying it out." Stephen Covey

Project Plan Contents

The Project Plan is the one Document that provides information regarding every testing activity from the Planning stages through to the completion of the Project. For Medium to Large Projects, the information that is contained in the Test Project Plan is usually prepared at a detailed level for the Test Section Managers, and summary information is available for the Test Manager.

If an automated Project Planning tool is used, as is normally the case, it can be used to facilitate the Planning activities, and as a monitoring mechanism during the Execution and Controlling stages.

A Test Project Plan should include the Tasks that were identified in the:

- Summary Test Plan
- Detailed Test Plans
- Test Calendar

And the Test Project Plan should have the following major sections:

- Write Summary Test Plan Definitions Sections (Chapter VI)
    - Definition of the Project

- Assumptions
- Test Scope
- Test Phases
- Test Calendar
- Write Summary Test Plan Action Sections (Chapter VI)
  - Dependency Projects Plan
  - Databases and Environment Plan
  - Human Resource Plan
  - Automation Plan
  - Quality Assurance Plan
  - Other Resources Plan
  - Risk Management Plan
  - Communications Plan
  - Budget Plan
  - Review and Approval Plan
- Write Detail Test Plan for each Phase (Chapter VII)
  - Introduction
  - Test Scope
  - Promotion Requirements
  - Test Calendar
- Prepare Test Material for each Phase (Chapter IX and X)
  - Data Sources
  - Test Documentation
- Prepare Test Environment for each Phase (Chapter IX and X)
- Execute and Verify Tests for each Phase (Chapter IX and X)
  - Execution Checklist
  - Execute the Tests
    - Test the Tests
    - Execute/Control
    - MIS
    - Adjust
- Control each Phase
  - Tasks
  - Quality
  - Scope
  - Budget
  - Risks

- Communications
- Closing (Chapter XIV)
  - Prior to Implementation
  - Post Implementation

Test Project Plan Reporting

In general, automated Project tracking tools provide the manager with the ability to identify and group Tasks, map resources against Tasks, do resource leveling, input cost information, and track the progress of the Tasks, resources and costs once testing begins. Additionally, they also provide a management methodology, and in some cases, ad hoc Reporting capabilities. Some of the more frequently used Standard Views (or Sorts) inherent within these tools are:

- By Milestone
- In-Progress Tasks
- Late Tasks
- Tasks to Begin Within a Specified Timeframe
- By Resource

Each of these Views provides information on the Test Status from a different perspective, and can help the Test Manager understand those activities that are either already off track, or that might be in jeopardy of getting off track.

By Milestone

If the Project Plan has been set up to identify key milestones, collapsing subordinate Tasks and displaying just the key events will result in a Summary Report. This will help the Test Manager maintain their perspective on the bigger picture, as it is very easy to get lost in the details when looking at a Test Project Plan with several thousand Tasks.

In-Progress Tasks

This View allows the Test Manager to focus on the Tasks that are underway. It is these Tasks that must be actively managed to ensure they stay on track. One helpful (although not fool-proof) indicator usually provided on this type of Report is the Percent Complete statistic. This percentage is normally manually estimated and calculated and is therefore subject to human error; however, it should be accurate enough to give the Test Manager some sense of whether the Task is in trouble or not. If, for example, a Task that had a ten-day duration is only 50% complete with one day left, there should be some doubt about whether it will be completed on time.

Late Tasks

Hopefully, after reviewing the In-Progress Tasks, the Test Manager will have some indication regarding whether or not some of them might subsequently show up on a Last Tasks Report. If not, this Report enables the Test Manager to see those Tasks that have missed either

> **Authors' note:** Keep in mind that once this information is available, the Test Manager needs to:
> - Explore the reasons for the problem
> - Determine what can be done to correct the underlying issue
> - Determine what adjustments could be made to get the Project back on track

their start dates or their completion dates. Once either Condition has been identified, an analysis must be done to determine the impact, and determine the possible adjustments that should be made if the late Tasks were on the critical path.

Tasks to Begin During a Specified Time

This View typically allows the Test Manager to get a sense of the Tasks that are scheduled to begin soon. Often, this Report is created to see the universe of Tasks for the next week and/or month. This helps the Test Manager ensure that all of the needed resources will be in place to handle the planned work.

By Resource

This View allows the Test Manager to get an overall sense of how the Tasks that are owned by any one resource are progressing. Some of the valuable Sorts within this View are:

- Tasks starting late
- Tasks completing late
- Tasks scheduled to start within a specified timeframe

The Test Manager should review these Sorts with some frequency to always have a general sense of how any one of the individual Testers is doing.

Test Project Plan

The Project Plan that was started in the Summary Test Plan can be expanded to include additional Tasks, such as:

Test Calendar

As the Test Procedures are written, the Test Days that will be executed for each Test Phase are identified, and are added to the Test Calendar. The Test Project Plan evolves along with the Test Calendar.

Other Testing-Related Tasks

Any other Tasks that were identified in either the Master Test Plan or the Detailed Test Plans should be included in the Project Plan.

After adding this information, the Test Project Plan could now look as follows:

	TEST PROJECT PLAN						
	Task Name	Planned Start	Planned Finish	Actual Start	Actual Finish	Predecessor Task	Assigned To
1	**Write Summary Test Plan**	7/01	7/31				
2	Definition	7.01	7/10				Test Mgr
10	Automation Plan	7/15	7/31				Dev Mgr
12	Other Resource Requirements	7/22	7/31				Test Mgr
13	**Write Detailed Test Plans**	8/01	9/13			3	
14	Unit Test Plan	8/01	8/07			4	Dev Mgr
28	Conversion Test Plan	11/01	11/15			27	End User
29	**Prepare Test Material**	7/29	8/30				
33	Functional Test	8/16	9/4			24, 32	
34	Extend Test Cases	8/16	8/22				Test Team
35	Prepare Test Procedures	8/22	8/30			34	
36	Test Date: 10/31	8/22	8/24			34	Test Team
37	Test Date: 11/1	8/23	8/25			36	Test Team
38	Test Date: 11/2	8/24	8/26			37	Test Team
39	Test Date: 11/30	8/25	8/28			38	Test Team
40	Test Date: 12/31	8/26	8/30			39	Test Team
47	**Prepare Test Environment**	8/05	9/03				IT
48	Copy Production Database	8/04	8/26				IT
49	Input Set-up Static Data	8/15	8/28			48	Test Team
50	Create Dynamic Data	8/20	9/03			48	Test Team
51	Prepare for Automation	8/15	8/30				Test Team
52	Implementation of Test Tool	8/15	8/22				IT
53	Prepare Test Cases for Tool	8/15	8/30				IT
58	**Test Execution**	9/01	11/08				
59	Unit Test	8/01	8/15				Dev Mgr
76	Day Five: 12/31	10/7	10/11				
77	Execution	10/7	10/09			73	Test Team
78	Verification	10/10	10/11			75	Test Team
79	Business Integration	10/14	10/25				Test Team
80	User Acceptance	10/28	11/08				End User

Figure 56 – Sample Expanded Project Plan

Summary

This chapter covered the following key points:
- The Project Plan consists of information from three primary sources:

- Summary Test Plan
- Detailed Test Plan
- Test Calendar
• The Project Plan should have the following major sections:
  - Summary Test Plan Definitions
  - Summary Test Plan Actions
  - Detailed Test Plan
  - Prepare Test Material
  - Prepare Test Environment
  - Execute and Verify Tests
  - Control Activities
  - Closing Activities

This sample Test Project Plan described in this chapter mentions the other Test Types, but it only covers the Functional Test Phase. Each Test Phase should have similar detail that would be added as the Detailed Test Plans are written and as more information becomes available.

As with the rest of the Detailed Test Plan, the Project Plan is a living Document and should be maintained throughout the testing effort to keep track of the Tasks and to manage the progress of the Execution. The Test Manager and Project Management Team may need this information presented in many different ways, but the key point for the Test Manager is to have a current Project Plan with the Tasks, dates and responsible parties identified so that the Task Status can be understood quickly and easily.

IX. BUSINESS FUNCTIONAL TESTING PROCESS

Testing an Application, regardless of its size or complexity, consists of two major groups of tests. When combined, these two groups help to ensure that the Application is properly exercised. Typically, both of these test groups are prepared and Executed very systematically, and are written to cover as many Conditions as possible within the time and Budget constraints of the Project.

- The first group of tests attempts to cover a wide range of non-functional Conditions that could possibly be encountered. These common testing Conditions form the basis for the Technical Testing Scenarios that are used in the Business Technical Tests, and are discussed in Chapter X with a primary focus on the following four Application Components:
    - User Interfaces
    - Reports
    - Files and Databases
    - Application Interfaces

> **Authors' note:** Functional Testing is based upon specific Business Scenarios, while Business Technical Testing includes Common Testing Conditions that can affect every type of financial Application.

- The second group is based upon functionally oriented Business Scenarios that reflect how the business works. The Test Team uses these scenarios to write Test Cases that are intended to simulate exactly how the Application will be used in Production. This group of tests is very specific to the Business Functions that are supported by the Application being tested. The creation of these tests is discussed in this chapter.

The Business Functional Testing Process consists of the following four activities:

- Test Material Preparation
- Database/Environment Preparation
- Test Execution
- Issues Tracking

Each of these four activities is discussed in the rest of this chapter.

Test Preparation

As the Master Test Plan is being assembled, other people on the Test Team can begin writing the tests that will be performed in the various Test Phases. These tests can be prepared using different types of data.

Data Sources

The data that is used for Functionally Oriented Testing can be obtained from two primary sources:

- Manufactured Data[34], including examples of cancellations, reversals and adjustments that cover:
    - Valid transactions with acceptable data
    - Valid transactions with incorrect data that should not pass edits
    - Invalid transactions with correct data
    - Invalid transactions with incorrect data

---

[34] A comparison of Manufactured Data versus Production Data is included in Chapter VI.

- Production Data that includes real-life examples of transactions, including:
  - Processed transactions of all types
  - Rejected transactions
  - Backdated transactions, etc.

This chapter will identify the uses of both types of data in creating tests.

Manufactured Data

The structure of functionally oriented tests consists of the following:

STRUCTURE		DESCRIPTION	EXAMPLE
1.0	Function	Functions are defined in the Summary Test Plan, and are usually related to Transaction Types in an Application	Buy Security
1.1	Business Scenario	Scenarios are a sub-set of the Functions, based upon some processing differentiation	Buy Equity
1.1.1	Test Case	Test Cases are extensions of Scenarios, based upon various processing Conditions that are defined in the Detailed Test Plans	Buy Equity under certain Conditions
1.1.1.1	Test Procedure	A Test Procedure is the lowest level of test, and is based upon an additional level of processing Conditions	Buy Equity under additional Conditions
1.1.1.1a	Description	Activities required to process the Test Transaction:	• Accessing the security Master File • Increasing the number of shares purchased (Trade Date) • Reconciling with the seller (Trade Date) • Increasing the number of shares held (Settlement Date) • Decreasing cash (Settlement Date) • Reconciling with the DTCC (Settlement Date)
1.1.1.1b	Test Transaction	Data Elements that are required to enter the transaction	• Account: 123456 • Security: ABCDEF • Transaction Type: Buy • Units: 100 • Price: $2.50
1.1.1.1c	Test Script	How to enter the transaction either manually or via some automated tool	• Go to Transaction Entry Screen • Enter Data • OK

1.1.1.1d	Expected Results	Results expected from this transaction, including where to look and what to look for	• Verify Transaction by:
			   – Viewing Account Holdings Screen for Account 123456
			• Account Positions should show:
			   – Units: 100
			   – Cost: $250.00
			   – Market Value: $250.00

1.1.1.1e	Test Log	Results obtained for this transaction, including supporting Documentation	• Completed Log • Screen Prints of Transaction Entry • Screen Prints of Account Holdings Screens

Figure 57 - Example of Test Hierarchy

This test hierarchy structure is used in the rest of this book.

Test Structure – Production Data

Using Production Data provides the Test Manger with a set of transactions that duplicate what can be expected when the new Application or Change moves into Production. There are two ways that Production Data can be used:

- The complete Production transaction can be used to initiate a test transaction, and the previously obtained results can be used as the Expected Results.

- The Components of the Production transaction can be used as a source to construct manufactured data. For example, the securities that were seen in Production can be used in Test Cases since the Testers can be sure that the security has been properly set up in the Security Master File.

There are several advantages and disadvantages to using Production Data as discussed previously.

Business Functional Tests are prepared using information that is defined in the Summary Test Plan and the Detailed Test Plans. The first step is converting Functions into Business Scenarios.

Business Scenarios

The process of writing functionally oriented Tests can begin as soon as the Functions have been identified in the Summary Test Plan, and the Testing Scope has been established. There should be close co-ordination between the people who write the Master Test Plan and the people creating the tests so that the tests support the Master Test Plan.

The following approach can be used to list all of the Business Functions and relate them to different sets of Conditions in order to create Business Scenarios. For instance, different types of instruments have different processing requirements and therefore each Function would have different Scenarios for each instrument.

	BUSINESS SCENARIOS [35]							
FUNCTIONS	EQUITY	CORPORATE BONDS	MORTGAGE BACKED SECURITIES	GOVERNMENT BONDS	MUNICIPAL BONDS	...	OTHER SECURITY TYPES
F-3.0 Buy Security	F-3.1	F-3.2	F-3.3	F-3.4	F-3.5	-	F-3.6
F-4.0 Sell Security	F-4.1	F-4.2	F-4.3	F-4.4	F-4.5	-	F-4.6
F-5.0 Cash	F-5.1	F-5.2	F-5.3	F-5.4	F-5.5	-	F-5.6

Figure 58 – Sample Business Scenario Matrix

Once these various Business Scenarios have been defined, they can be identified for use in several Test Types, including:

- Functional Test
- Business Integration Test
- User Acceptance Test
- Parallel Test
- Regression Tests

The development of functionally oriented tests is covered in this chapter.

Test Case

A Test Case is a sub-set of a Business Scenario, which involves specific processing Conditions or limitations. Some of these Conditions and limitations can be:

---

[35] This is a sample of the Functions that could be covered in an Application.

Processing Conditions

When specific data has processing implications, a Test Case should be written for each of the Conditions[36] that create a separate processing path. For instance, each Business Scenario can also be extended by other Conditions and result in multiple Test Cases.

Processing Limitations

In cases where the processing of the Application is limited, the Test Cases should be written to test each of the limitations, such as if the Reporting Application does not allow weekly or bi-weekly Reports, there should be Test Cases for each of these as well as for the available frequencies.

The Business Scenarios can be extended by applying the Conditions to each Scenario to create the Test Cases as shown in the following example.

TEST CASE MATRIX – F-3.0 BUY SECURITY						
CONDITIONS	ADD	MODIFY	VALIDATE	REJECT	CANCEL	REVERSE
F-3.1 Equity	3.1.1	3.1.2	3.1.3	3.1.4	3.1.5	3.1.6
F-3.2 Corporate Bonds	3.2.1	3.2.2	3.2.3	3.2.4	3.2.5	3.2.6
F-3.3 Mortgage Backed Securities	3.3.1	3.3.2	3.3.3	3.3.4	3.3.5	3.3.6
F-3.4 Government Bonds	3.4.1	3.4.2	3.4.3	3.4.4	3.4.5	3.4.6
F-3.5 Municipal Bonds	3.5.1	3.5.2	3.5.3	3.5.4	3.5.5	3.5.6

Figure 59 – Sample Test Case Matrix

In creating the Test Cases, there should be a balance between the need for detailed testing with the need for Project efficiency. Creating reams of paper that document all of the Test Cases does not result in an efficient testing Project. However, in most organizations, time is of the essence in the testing process, so all of the Test Cases should be reviewed with a number of questions in mind:

- Is this Test Case too comprehensive or too limited?
- Could this be combined with another Test Case?
- Is this Test Case redundant?

Test Procedure

The Test Procedure defines the series of transactions that are required to implement one aspect of the Test Case. By applying additional Conditions, the Test Cases can be further extended to establish the Test Procedures.

TEST PROCEDURE MATRIX – F-3.1 BUY SECURITY – EQUITY							
CONDITIONS	SUFFICIENT UNITS	INSUFFICIENT UNITS	LATE TRADE	...	...	...	...
F-3.1.1 Add	3.1.1.1	3.1.1.2	3.1.1.3				
F-3.1.2 Modify	3.1.2.1	3.1.2.2	3.1.2.3				
F-3.1.3 Validate	3.1.3.1	3.1.3.2	3.1.3.3				

Figure 60 – Sample Test Procedure Matrix

---

[36] A list of Conditions is included in Chapter VII.

There are five detailed sections in a Test Procedure:

Test Objective

The objective for each Test Procedures should be defined. For instance, the purchase of U.S. equities involves several transactions, including:

- Accessing the security Master File
- Increasing the number of shares purchased (Trade Date)
- Reconciling with the seller (Trade Date)
- Increasing the number of shares held (Settlement Date)
- Decreasing cash (Settlement Date)
- Reconciling with the DTCC (Settlement Date)

Test Transaction

The Test Transaction consists of the Data Elements that are needed to input the transaction. Transactions may be simple or complex. A simple transaction may merely require a list of specific Data Elements, while a complex transaction might require that if a certain Message is received while the transaction is being entered the Tester must go to another Screen.

Test Script

The Test Script defines how the Test Transaction should be entered.

- If the test is manual, the Test Script identifies the Screen(s) that should be used and any unique information that might be required to actually enter the transaction.
- If the test is automated, the Test Script is the automated script itself.

Expected Results

Each Test Transaction must have Expected Results. The results will specify the responses that are expected in the:

- User Interface
- Other Screens in the Application
- Reports
- Application Interfaces
- Data Storage

It is possible that one transaction will create a series of Expected Results that will impact several other Applications through Interfaces. This should be defined before the tests begin.

Test Log

The Test Log records the results of the test, and includes Actual Results, as well as any supporting Documentation, such as Screen Prints, Reports, etc., so that the Developer can see the Defect and recreate the test. It is important to make these Documents available to the Developers and to maintain these copies for the Auditors' review. The best solution is to store the Screen images electronically and to scan in paper so that the entire history of the test is available in electronic form.

Some of the available Test Tools that are discussed in Chapter XV can support this requirement, and other firms use programs such as Lotus Notes.

Examples of this hierarchy for Functionally Oriented Tests are shown in the following chart.

	SECURITIES	BANKING	INSURANCE
Function	Sell Security	Electronic Funds Transfer	Purchase of an Insurance Policy
Business Scenario	Sell Equity	Immediate Funds Transfer	Purchase of a Term Policy
Test Case	Sell Equity with insufficient units	Transfer for US Dollars to an account in a bank in another country	Purchase of a five year fixed term, single premium Insurance Policy
Test Procedure	Add Sale	Add Transfer	Add Purchase
Description	Sell 1,000 IBM At Market Account 12345657 500 shares available	Transfer $2,000.00 Account 2345678 Today To Bank ABC in GB SWIFT 887766 Account 789789789	Purchase Policy $100.000.00 5 Year Term 1st Year Receivable of $672.50
Transaction	List of the required Data Elements to initiate the transactions	List of the required Data Elements to initiate the transactions	List of the required Data Elements to initiate the transactions
Test Script	Document showing how to enter the transaction manually or through an automated tool	Document showing how to enter the transaction manually or through an automated tool	Document showing how to enter the transaction manually or through an automated tool
Expected Results	A definition of the expected Error Message	Location of where the tester can see the debit posted A Report that shows that the funds were sent	A copy of the notices to the insured and the beneficiary
Test Log	A record of the Actual Result and copies of supporting Documentation	A record of the Actual Result and copies of supporting Documentation	A record of the Actual Result and copies of supporting Documentation

Figure 61 - Example of Test Hierarchy

Business Scenario Documentation

Business Scenarios are related to the Functions that are identified in the Summary Test Plan and should be defined in the Detailed Test Plans in a series of Matrices that are usually part of the Scope section of these Plans.

Test Case Documentation

The Test Case Documentation should have the following information:

Test Case Identifier

This number should be unique and, depending on the naming convention, it can be used to identify the Test Types that are covered, what phase of testing will Execute the test, or the business line that is using the test.

Objective of the Test

This is the summary of the Conditions that are being tested, and should be cross-referenced to the test Matrix that is included in this Test Cycle. Depending on how much cross-referencing an organization needs, the Audit

Department may also require a link between the Functional Requirements or Design Documents and the Test Cases.

This level of information requires that all of the Requirements and Design Documents have a naming or numbering convention that will always allow a quick cross reference to the Test Case. Similarly, the test Matrices that are created should have unique numbers for the Conditions that are being tested. The Test Procedures will provide the detailed procedures and Expected Results for executing the tests, so this section only needs to identify the Conditions at a summary level.

One structure could include identifiers such as:

- Test Case Number
- Document Cross-Reference (Document name, version and page number)
- Retest Number

Input Conditions

This section should list the specific input types that are being tested, such as the Screens, Files or Messages with some identifiers to distinguish between one screen and another, one file, etc. Again, this section should be cross-referenced to the Matrices defined in the Detailed Test Plan.

Output Conditions

As with the input, this section should list the format, and the Conditions covered by the output that is being tested, such as Screens, Reports, Interfaces or Messages with some identifiers to distinguish between one screen and another, one file, etc.

Environmental Needs

This section should list the data that is needed in order to execute this test. This should include some of the following:

- Test Calendar considerations
- Static data set-up requirements
- Dynamic data requirements
- Automated Tools for input or output

Date Requirements

There may be specific Requirements regarding the date or type of date that is required to execute the Test Case. For example, a test of the Client Statement processing will require a month-end date, but may also require a number of business dates following month-end to cover Conditions such as late trades, reversals, etc. This section should cover all dates that are required for the Test Case.

Inter-Case Dependencies

In some cases, a particular Test Case cannot be executed until some prior test or Function has been tested or at least executed. In the best-case scenario, if the Test Team is testing the format and data on a Client Statement, the Screens and Functions that allow the Client Statement to be requested already should have been tested.

Test Case Documentation

The following is a sample of a Test Case format:

TEST CASE NUMBER	TEST CASE NAME	TEST DAY REQUIREMENTS	DEPENDENCIES	SCREENS REQUIRED
	Field Editing of Trade Entry Screen	One overnight batch needed	None	

INPUT CONDITIONS	CURRENCIES NEEDED	ACCOUNTS NEEDED	SECURITIES NEEDED	OUTPUT CONDITIONS
Test Trade Entry Screen for all Transaction Types	USD	123456 234567	US Equity	Test Daily Trade Activity Report for all Transaction Types
Test Required and Optional fields on Trade Entry Screen			US Corporate Bonds	Test Matrix Coverage: R1.1 B and D, R2.1-10 A-E
Test Matrix Coverage: S1.2 C-F, S2.1-9 A-G				

Figure 62 – Sample Test Case

Keeping track of the coverage of the Test Cases is also very important to the Tester who is writing the Test Procedures, and these Matrices make it clear what needs to be tested. Of course, it is important to remember that all of these Documents are dynamic in nature and one of the great challenges in software testing is to keep them current.

Test Procedure Documentation

From the Tester's perspective, the Test Procedure is the foundation of the testing effort. This Document defines what they are to test and identifies the Expected Results. The Test Procedure Document should include the following sections:

Objective

The Objective of each Test Case should be identified.

Test Case Identifier

There should always be a link to the Test Case by including the Test Case ID on the Test Procedure.

Test Procedure Identifier

As with the Test Case ID, this number should be unique and the naming convention can be used to give more information regarding the nature of the Test Procedure.

Test Coverage

The Test Case documents, at a summary level, the Conditions that are being tested. This section of the Test Procedure should identify the specific Conditions that are being tested, such as Transaction Types, security types, Screen Functions or Report Conditions.

Test Transaction

The actual Test Transaction consists of a series of Data Elements that must be entered into the Application to perform the test.

As previously mentioned, the data that is used for Business Functional Testing can be obtained from two primary sources:

- Manufactured Data, as well as examples of cancellations, reversals and adjustments, and which includes:
  - Valid transactions with acceptable data
  - Valid transactions with incorrect data that should not pass edits
  - Invalid transactions with correct data
  - Invalid transactions with incorrect data
- Production Data that includes real-life examples of transactions, including:
  - Processed transactions or all types
  - Rejected transactions

The data that is used should be clearly documented in the same format that will be needed for data entry, either manual or automated, to speed the Execution of the Test Procedure.

> Authors' note: Tests should be designed with the Testers' skill level in mind, while remembering that less experienced testers cost far less.

Test Script

This section is where the Test Manager needs to make an early decision on how the Test Procedure will be used. If the Test Procedure is going to be used by junior level Testers or End Users who are not completely familiar with the Application, the procedural steps should be documented in greater detail in order to serve as a Training tool as well as a testing tool. This may include step-by-step instructions for Screen log-on, navigation, etc.

With a more seasoned team, the Documentation can be reduced. An experienced person may only need the procedures Document in order to know what Screens should be used.

Expected Results

As with the procedural steps, the level of detail in this section can vary according to the experience of the Test Team. Junior level Testers may need to have detailed Documentation on what Screens should appear, what Messages are anticipated, etc., for a successful test. The more senior Tester may only need the Expected Results where there is some calculation being done in the presentation of the data.

Test Log

Actual Results

The Actual Results may need to be documented in detail or it may be that the Pass Status will indicate that the Actual Results matched the Expected Results. This is an area where comments can be documented if the Expected Results may have been incorrect, either based on the Design specification or from a business standpoint.

Pass/Fail Status

If the Procedure is also being used as a Test Log, the Status should be updated once the test has been run through to completion.

Documentation

As the tests are conducted, the Tester should capture the Documentation that proves the test either passed or failed.

- If it passed, then it is needed to show the Auditors (and the End Users) that a specific activity was tested and works.
- If the test failed, the Documentation is needed to show the Developers what happened so they can correct the situation.

Issues Tracking Numbers

When a Defect has been uncovered while executing the Test Procedure, the Issues Tracking Number should be documented so that the Re-test of the Test Procedure is tied to the resolution of the Issue. The Issues Tracking process is discussed later in this chapter.

Version of Code

It is always important to know and document the version of the Code that the Test Procedure was executed against.

The Test Manager has considerable flexibility when defining the level of detail required in the Test Procedure, but the decisions regarding what should be included should be made early in the Test Planning process, since the time spent on preparing Test Procedures will change based on the level of detail.

TEST NUMBER	TEST NAME	TEST DATE	VERSION	TEST OWNER
	Field Editing of Trade Entry Screen	10/31	2.04	Tess Tester

PROCEDURE	EXPECTED RESULT	ACTUAL RESULTS	PASS/FAIL	ISSUE #
Log on to Application				
Enter Function: TRADE				
Enter Account: 123456	Message for invalid account #	Message box appears	P	
Enter Account: ABC123	Account name should appear	ABC Client appears in account name field	P	
More steps would follow				

Figure 63 – Sample Test Procedure

The Test Procedure can continue to be used after it has been executed. Depending on the resources and future Plans, it may make sense to build a Test Library of all of the Test Cases and Test Procedures for future use. This can substantially reduce the preparation time for Regression and Stress Testing. With this in mind, the Test Procedures must be kept current so that they properly reflect how the Application works as Changes are incorporated.

Database/Environment Preparation

When executing the tests, the Testers will need access to many different categories of information that may include:

- Accounts
- Securities
- Positions
- Counterparty Files
- Broker Files
- Security Access Permissions
- FX Rates
- Currency Codes
- Country Codes
- Factors
- Corporate Action Announcements, etc.

> **Authors' note:** If the Environmental set-up is incomplete or incorrect, the testing efforts will be seriously affected. Do not try to skip this step or rush it.

This information is usually held in the Files or Database in the new Application or in other existing Applications that must be accessed. The information may be copied from the old Application, or an entirely new File may have to be created and populated. In either case, the new Files should be examined and tested on their own before the actual testing begins.

For example, to execute the Test Transaction defined earlier in this chapter, the following Files will have to be properly loaded before the test can occur:

ACTIVITY	FILE NAME	PURPOSE OF FILE FOR THIS TEST
Log-on to the Application	Security Access	Permission to Log-on
Access the Trade Screen	Security Access	Permission to use the Screen
Enter an Account Number	Customer Master File	Determine if the account is valid
See Error Message	Warning/Error Message File	Relate an explanation of the error to the activity

Figure 64 - Files Needed for Test Example

The Test Manager must work with the Developers or the vendor to identify what must be set up and what parameters must be established for each File in order for the set-up to be correct.

TEST NUMBER	TEST NAME	TEST TYPE	TEST REGION	TEST OWNER
F-1.1	Set-up Account	Functional	Test 1	Elizabeth Owner

ACCOUNTS COVERED	PROCESS REQUIRED	SECURITIES REQUIRED	POSITIONS REQUIRED	TEST PROCESSING
123456 234567 345678	Enter into Database	Copy of Production Securities Database as of 6/30/2002	For covered accounts from Production as of 6/30/2002	Load using QARun scripts

Figure 65 – Sample Extract from Functional Test Plan

If the Master Test Plan calls for multiple Test Regions, it may be necessary to establish these Files in more than one region. Since it is normal for these Files to change to some degree throughout the testing process, it is possible that the Files in one region are not the same as the Files in another region. There may be a good reason for this, but it has to be closely managed.

Test Execution

Once all of the Planning has been completed, the tests prepared and the Environment established, the Execution Checklist can be established and the actual testing could begin. There are three parts to Test Execution:

Test Transaction Entry

Tests can be entered in three ways:

Manual Entry

Manual entry is valid for Projects that require a small number of Test Transactions, where the cost and overhead associated with automated testing is unnecessary. Manual testing usually involves manual test tracking and manual Documentation

Automated Entry

Automated entry is useful for Projects that have a large number of repetitive tests, where it is worthwhile to create some form of test automation. Automated entry of test data may involve manual test tracking and manual Documentation or some level of automation.

Test Automation is discussed in Chapter XIII.

Combination of Manual and Automation

Most Projects have a combination of low volume transactions and high volume transactions. In this case, the Test Manager should review the number of tests that affect each Transaction Type and decide which transactions should be automated.

Test Execution Documentation

The Document that identifies the results of the Test Execution is the Test Log, which was described earlier in this chapter. The Test Log records the results of the tests, including:

- Tests passed successfully
- Tests differed to a later day
- Tests that did not pass

Test Verification

For every test there should be a verification step. The verification or validation may be accomplished by reviewing reports, screens, databases or file outputs. The method for verifying the test results is defined in the Test Procedure as the Expected Results.

Issues Tracking

The Defects that are identified when a test's Expected Results were not obtained are recorded in the Issues Tracking Report.

Issues Management

The purpose of the Issues Tracking Report is to record Defects and track their resolution.

Objectives

The objective of the Issues Tracking process is to provide a uniform method for tracking and reporting issues that arise during the testing process. This process is the mechanism for all of the interested parties to monitor the open and closed issues in order to assess the progress of the Project and the stability of the Application.

> **Authors' note:** Once a testing effort has started, the Issues Tracking Report will be accessed continuously.

Issue Tracking Procedures

The procedures concerning Issues Tracking must be clearly defined and understood by the Testers, Developers and End Users. There are many ways to handle the Issues, which depend on the Project, the experience of the Testers and the involvement of the End Users. Some questions that need to be answered in the procedures are:

Who Enters the Issues?

The organization may have the Testers enter their own Issues, or there may be a central person. If the Test Team is made up of more junior level people, the team may have a person who is responsible to reproduce the Issue before it is formally logged.

Who Assigns the Severity Level?

Ultimately, the End User should review all of the Issues to assign the Severity level, but in practice, the Tester usually assigns some of the Severity levels. It may be that all new Issues are assigned a Severity of "New" and then a review team assigns the true Severity based on a more careful consideration of the Issue. Defining this procedure early in the Planning process will allow the Issues to be logged in a timely manner and not require anyone to make Severity decisions without time to consider the impact of the Defect.

Who Can Close an Issue?

Again, according to the experience level of the Testers, a Test Team Section Leader or the Test Manager may choose to review all of the Issues before they are closed to make sure they have been properly Re-tested. It may also be that the End User wants to participate in the Issues resolution by signing off on all closed Issues.

While the procedures are important, if there are too many steps in the logging and resolution process, a bottleneck could be created. As always, the testing is the most important Task, and the Test Team should not be overwhelmed with paperwork and procedures.

> **Authors' note:** It's just business.

Inevitably the Issues Tracking process becomes an area of potential conflict in the Project. The Test Manager must ensure that the integrity of the testing and the Issues are Maintained, and that the human factors do not compromise the process. Remember, it is nothing personal!

Adjust the Plans

Do not establish a process and expect it to work automatically. The Issues Tracking process must be tailored to each company's unique needs and the participants' personalities, and it must be monitored throughout the Project to ensure that nothing has changed that would require process modifications.

Issues Documentation

The reporting and tracking of the Defects that are found during Development is a key component of the testing activities. However, there can be many difficulties in the process of creating and updating the list of Defects, more commonly called the Issues Tracking Report.

Depending on the organization, the Issues Tracking process can be used to keep track of some, or all, of the following types of Issues:

> **Authors' note:** By relating the report to Issues rather than Defects, the Test Manager can hopefully avoid at least one of the inter-organizational difficulties: the incorrect assumption that all of the items that are tracked are Code Defects or Defects in the Development process.

- Design Defects
- Design Changes or Enhancements
- Program Defects
- Testing Defects
- Management Issues
- Environment Issues
- Questions

At the beginning of the Project, the Project Management Team should decide upon the types of Issues that will be included in the Issues Tracking Report, and the types of Issues that will be reported to various levels of management and interested parties. It may not be necessary to report on all of these types of Issues to everyone.

Depending on the types of Issues that are logged, the form that is used to record the Issues should reflect all of the required information. In order to properly track the problems that arise from testing, the following information should be on the form when the Issue is originally documented:

Problem Number (Issue Tracking Number)

The Issues Tracking number should always be unique; however, the numbering scheme may be designed to identify the type of problem, or the place where it was uncovered, such as one numbering scheme for Production Issues, and another for Development.

The structure for the Issues Tracking Number could include:

- Application Name (If the Issues Tracking Report is used by more than one Test Team)
- SDLC Phase in which the Issue is found
- Date the Issue was identified

- Sequence Number for the Issue on that day

Application

Because multiple Applications can use the same Issues Tracking Report, it is important to identify the Application where the Issue was uncovered.

Function

The Function that is being tested should be identified in the Issue Tracking Report. The Test Manager should determine the granularity of the items that are recorded. The list may correspond to Business Functions, but for ease of entry and reporting, the number of selections should be limited.

Version

The Version of the Code that was in place at the time the Issue was discovered must be entered for reporting purposes and to help the Development Team understand the possible source of the Issue.

Database

If there are multiple Test Regions, the data source or the Database configuration may be different. Identifying the Database is essential to reproduce the problem and for the Re-test.

Short Description of Problem

In most cases, it is not practical to use the whole description of the Issue on all of the Reports, so a Short Description is used. The Test Manager should provide Guidelines for the short description that will help managers quickly understand the Issue.

Long Description with Attachments

This section should include a detailed description of the Issue, the steps that led to the Issue, as well as the Expected Result and Actual Result. In addition, there should always be a place or a mechanism to attach Screen Prints, Error Messages, Reports, and other Documentation that illustrates the Issue's Scope.

Test Case/Procedure Number

It is very helpful when reproducing or Re-testing an Issue to have the Test Procedure Number(s) that were executed in uncovering the Defect.

Is The Defect Reproducible?

Depending on the procedures regarding the logging of Issues, someone may be assigned to reproduce a problem before it goes to the Developer for analysis.

Found By

When the name of the person who found the Defect is on the form, the Developer can easily identify the person who knows the most about it. Some companies do not want Developers and Testers to talk directly, preferring that all communications go through a central point; however, the Tester's name should still be recorded.

Status

The Status field on the Issue Tracking form can be a contentious area. In most cases the Status is used to track the progression of the Issue, so the valid Status types should be at the level of detail that will be needed for Reporting. Some of the Status levels could be:

- Open
- Closed
- Re-test
- Re-Open
- Ready for Production
- Deferred
- Enhancement

The procedure may be that an Issue goes from Open to Re-test to Ready for Production to Closed. There are many different progressions that an Issue may follow, but the valid Status types should reflect all of the possible paths.

The best-case scenario for progression of an Issue would be:

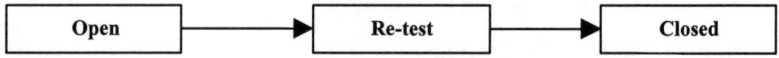

But these scenarios also (and more likely) happen:

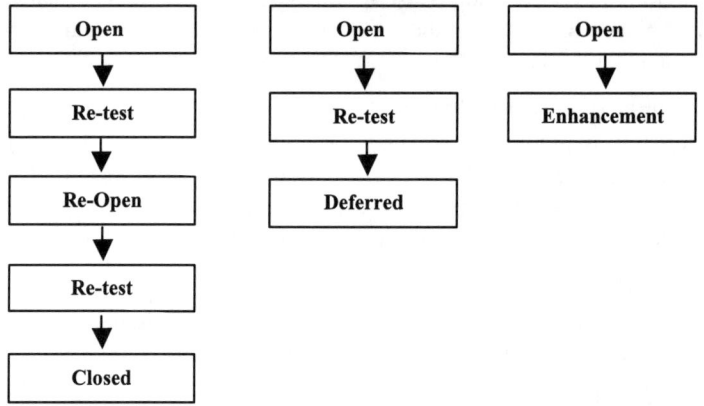

Figure 66 – Possible Testing Processes

The second area of concern is to identify who should be allowed to make changes to the Status of a given Issue. Again, this is a procedural decision that should be made early in the Project before any Issues have been recorded.

- In some organizations, only the managers can change the Status. This can create a bottleneck in the processing of Issues.
- Other organizations allow all Testers and Developers to make updates. This can reduce Control.

While the procedural decision should be made early in the Project, it can change to meet the evolving needs of the Project Team.

Severity

Defining the Severity also can cause reporting problems. While on paper the Severity definitions seem very clear, in practice they can be easily misinterpreted. It may be that the Tester who records the Issue sees it as having a serious business implication while the End User may see the Issue as less serious because they know of a potential work-around.

Some of the Severity levels that are often used are:

- Critical
- High
- Medium
- Low

> **Authors' note:** The Severity level that is assigned should reflect the End Users' View, not the Testers.

The Severity level may be used to determine the cut-off for Issues that will be fixed prior to Production. For instance, if all of the Issues can not be resolved prior to the live date, the Project Team may decide to focus on the Critical or High Issues and defer the Medium and Low Issues to a future release. It is important that the Severity reflects the End Users' opinion and not the Tester's.

Throughout the Project, the End Users should prioritize the Issues within each category so the Development and Test Teams know which Issues should be addressed next.

Assigned To

This field will change as the Issue moves through the resolution process. The Issue may be assigned to the Developer when it is being fixed or the Tester when it is in Re-test Status, but it is always important to have a person assigned to every Issue at every point so management can get an update during its resolution.

> **Authors' note:** Issues without an owner do not get attention.

Resolution

Once the Issue has been recorded, analyzed and resolved, the Resolution section should be completed. This section should minimally have the following information:

Resolution Type

This field is a description of the actions taken to resolve the Issue, which may not always be a Programming Change. It could be a procedural revision by the End User, it could be a mapping modification, or it could be that it is a misunderstanding on the part of the Tester. In all cases, the Resolution Type should be documented in order to close the Issue.

Resolution Date

This date shows when the resolution was completed. In the case of a Programming Change, this is when the Developers made the repair, not when it was Re-tested.

Resolved By

Again, it is important to have a name associated with the resolution in the event there are questions or problems with the Changes that were made.

Re-Testing

In the case where the Issue required Programming Changes, the Issues Tracking Report should contain information on the Re-testing effort.

Re-test Status

This indicates if the Re-test passed or failed. A sequence number should be included to indicate the number of times the Issue has been retested.

Re-test By

This identifies the person who Re-tested the Issue. This is particularly important if the Re-test fails and the Developer needs to get more information on the Re-test.

Re-test Date

This helps to track the Issue through the resolution, especially in case the Issue needs more than one Re-test.

The forms that are created to create and track Issues may be as individual as the organization, but these procedures should always be designed to simplify the process of tracking the Issues and reporting them to all interested parties.

Summary

There are four parts to the Business Functional testing process:

- Test Material Preparation
- Environment Preparation
- Test Execution
- Issues Tracking

This chapter describes how tests are prepared, Executed and tracked. When writing the tests, the authors recommend using the following hierarchy of activities:

- Business Function
- Business Scenario
- Test Case
- Test Procedure
  - Description
  - Test Transaction
  - Test Script
  - Expected Results
- Test Log
  - Actual Results
  - Documentation

Each of the steps in preparing tests is discussed in detail.

After a test has been performed and an Issue has been identified, the Test Team should use an Issues Tracking Report to record and track each item.

X. BUSINESS TECHNICAL TESTING PROCESS

In Business Technical Testing, the Test Team uses combinations of data that are logically possible, but which are not derived directly from the Business and Functional Requirements Documents[37]. In general, this second group of tests focuses on the following system Components:

- User Interfaces
- Reports
- Files and Databases
- Application Interfaces

The Testing Process for Business Technical Testing consists of the same four activities that are used in the Process for Business Functional Testing:

- Test Material Preparation
- Database/Environment Preparation
- Test Execution
- Issues Tracking

This Process was covered in Chapter IX, and is not repeated here.

Business Technical Tests for these Components are conducted to validate that:

- What was delivered is what was expected
- Each component's basic activities operate as specified and are stable
- The Application can perform within a full range of Test Scenarios
  - In Positive Testing, the Test Scenarios represent what is expected to occur
  - In Negative Testing, the Test Scenarios represent what is not expected to happen, but could occur

Most of the Business Technical Tests should be conducted early in the Test Calendar to ensure that the Application is stable enough to process the more complex Business Scenarios. Normally, this level of testing occurs just before, or in conjunction with, the Functional Test.

Preparing Business Technical Tests

The Technical Scenarios, Conditions and Application Components should be identified in the Summary Test Plan and in the Detailed Test Plan. The list of the Business Technical Scenarios that can apply to the four Components includes the following general Conditions as well as some that are specific to a Component:

- Completeness
- Stability
- Navigation
- Appearance
- Field Edits
- Defaults
- Warning and Error Messages

---

[37] Business Functional Testing is discussed in Chapter IX.

- Security

Applying the Technical Conditions to the different elements of each Component results in a range of Technical Scenarios that could be identified in a set of matrices.

TECHNICAL SCENARIO MATRIX – USER INTERFACES								
SUB-COMPONENTS	COMPLETENESS	STABILITY	NAVIGATION	APPEARANCE	FIELD EDITS	DEFAULTS	ERROR HANDLING	
TS-1.0 Log-On	TS-1.1	TS-1.2	TS-1.3	TS-1.4	TS-1.5	TS-1.6	TS-1.7	
TS-3.0 Buy Security	TS-3.1	TS-3.2	TS-3.3	TS-3.4	TS-3.5	TS-3.6	TS-3.7	
TS-4.0 Sell Security	TS-4.1	TS-4.2	TS-4.3	TS-4.4	TS-4.5	TS-4.6	TS-4.7	

Figure 67 – Business Technical Scenarios Example

The previous Matrix shows the Testers the Technical Conditions that need to be covered for each of the Screens regarding navigation, editing, and data. The following Matrix shows Testers the Technical Conditions that need to be tested for one specific Screen. Each Screen (or groups of very similar Screens) would have a separate Matrix.

The Technical Scenarios that were identified in the Summary Test Plan can be extended to create Technical Test Cases as shown in the following chart.

TECHNICAL TEST CASE MATRIX - TS-3.0 BUY SECURITY SCREEN					
COMPONENTS	FIELD SIZE	REQUIRED	OPTIONAL	RESTRICTED	SPECIALIZED DATA TYPES
TS-3.1 Completeness					
TS-3.2 Stability					
TS-3.3 Navigation					
TS-3.4 Appearance					
TS-3.5 Field Edits	TS-3.5.1	TS-3.5.2	TS-3.5.3	TS-3.5.4	TS-3.5.5
TS-3.6 Defaults					
TS-3.7 Error Handling					

Figure 68 – Business Technical Test Case Example

Once the Test Cases have been defined, the Test Team can add another level of detailed Conditions to create the Test Procedures.

TECHNICAL TEST PROCEDURE MATRIX – TS-3.5 BUY SECURITY SCREEN FIELD EDITS						
	MAXIMUM	MAXIMUM – 1	MAXIMUM + 1	MINIMUM	MINIMUM – 1	MINIMUM + 1
TS-3.5.1 Field Size	TS-3.5.1.1	TS-3.5.1.2	TS-3.5.1.3	TS-3.5.1.4	TS-3.5.1.5	TS-3.5.1.6

Figure 69 – Business Technical Test Procedure Example

This process of documenting Business Technical Tests can be used for each Component, under the Conditions that are identified in the following sections.

User Interfaces (UI)

Testing Applications that use Graphical User Interfaces (GUI) is different from testing traditional Text-Based Applications, and therefore automating the testing process is also different. Text-Based User Interfaces usually follow a very logical and straightforward sequence of events, while GUI's typically have many different Components on each Screen that can allow the user to perform Tasks in a different sequence and to move rapidly from one area of the Application to another.

When Testers move from testing Text-Based Applications to GUI Applications, they need to recognize that they will have to use different techniques, and the process itself needs to be tested in more detail.

Testing a Text-Based or a Graphical User Interface requires both of the two major groups of tests; the validation of the Business Functions, and the testing of the Business Technical Conditions. The combination of these two efforts fully exercises the UI, and once the UI Screens have been stabilized, they can be used as a tool to support the testing of the rest of the Application throughout the entire Project.

This section reviews the common testing Conditions that should be used when testing the User Interface of any Application, whether graphical or text-based, and regardless of the size or complexity. The Technical Scenarios for UIs include:

- Completeness
- Stability
- Navigation
- Appearance
- Field Edits
- Defaults
- Warning and Error Messages
- Security

Authors' note: Test data should be established that will support more than one test category.

Test Preparation

To establish these tests, the Test Manager should prepare a series of Matrices to guide the writing of the Test Cases for the Business Technical Testing. These Matrices are typically assembled from a variety of sources that provide information on Screen layouts, navigation, field sizes, edit criteria, business rules, etc., and are used as Control Documents to prepare the tests.

It is up to the Test Manager to determine the best layout for these Matrices and the following table provides information on the type of source Documents that typically contain the information that is needed to prepare the Matrices.

TECHNICAL SCENARIO	POSSIBLE SOURCES OF INFORMATION
Completeness	Functional Requirements Document
	Technical Design Document
	GUI Layout/ Navigation/ Report Document
	Screen Mock Up Document
Stability	Technical Design Document
Navigation	Functional Requirements Document
	Technical Design Document
	Menu Map
	GUI Layout/ Navigation/ Report Document
Appearance	Screen Mock Up Document
	Corporate Standards Document
Field Edits	Data Design Document
	Data Dictionary
	GUI Layout/ Navigation/ Report Document
Defaults	Data Design Document
	Data Dictionary
	GUI Layout/ Navigation/ Report Document
Warning and Error Messages	Design Document
	GUI Layout/ Navigation/ Report Document
	Warning/ Error Message Listing
Security	Functional Requirements Document
	Corporate Security Guidelines

Figure 70 – Sources of Information

The location of this information will vary from organization to organization, and in some cases, the information that is needed is not formally documented at all. If these Documents do not exist or are incomplete, the Test Manager must obtain the necessary information before the Test Cases and Test Procedures can be created.

Authors' note: The circumstances for each Project are unique and the Test Manager must adapt to the situation.

If it is not formally documented, the Test Manager will likely have to interview a variety of people (Business Analysts, Developers, End Users, etc.) and document the results. The Test Manager must determine if this effort negatively impacts the overall Test Calendar, and ensure that the appropriate parties are aware of the reasons for any delay.

There should be a series of Test Cases to support each of the Technical Scenarios, and the Test Cases will require several Test Procedures to validate each case. The Test Cases and Test Procedures for the Technical Scenarios are written using the same process as for the functionally oriented tests.

The following Business Technical Scenarios must be essentially complete before the Functional Test begins:

- Completeness
- Stability
- Navigation

However, the remaining types of Conditions can be performed in conjunction with the Functional Test, and could run concurrently or serially depending on the staffing and the Plan, including:

- Appearance
- Field Edits
- Defaults
- Warning and Error Messages

All of these Technical Scenarios are described in the following sections.

Completeness

When the User Interface Modules are first delivered it is important to validate that everything that was scheduled for delivery was actually delivered, and as simple as this seems, many Testers frequently skip this step. When the upfront validation does not occur and subsequent problems are identified, valuable time could be wasted in testing incomplete UIs and Applications.

> **Authors' note:** "Trust, but verify." — Ronald Reagan

In many cases, time is wasted as Testers attempt to run thorough tests against software that has only been partially delivered, and/or that is too unstable to allow tests to run to completion. To avoid this, a simple checklist helps to ensure a good delivery. Very shortly after delivery the Test Team should know the answers to the following questions to determine if the delivery is complete:

TEST CASE	CONDITIONS
Validate System Completeness	Validate that all anticipated Modules have been delivered.
	Validate that all of the Screens within each Module have been delivered
	Validate that all of the menu options are available • File • Edit • View • Tools • Actions • Help, etc.
	Validate that all the expected Data Elements are accounted for on every Screen
	Validate that all of the defined boxes/buttons are included: • List boxes • Dialogue boxes • Check boxes • Radio buttons • Push buttons, etc.
	Validate that all of the Table Data is loaded
	Validate that all required graphics and animations are shown properly

Figure 71 — Sample Test Case for User Interface Completeness

If the UI module that was delivered was not what was in the Plan, the Test Manager must decide if it makes sense to continue with the rest of the testing. If, for example, the Application is so unstable that it crashes more than it is available, the Test Manager might decide to refuse the delivery until a more complete Module is ready. However, if the full universe of table data is not yet complete or every menu option is not available, the Test Manager probably would not delay further testing until the remaining activities are completed.

Once it has been decided that the UI is sufficiently complete, the stability and navigation tests should be performed.

Stability

The Module must be tested for stability before the Functional Test can begin. Some of the key questions to ask are:

TEST CASE	CONDITIONS
Validate System Stability	Validate that all table data is retrievable
	Validate that all of the basic activities can be performed: • Add • Change • Delete • Search, etc.
	Validate that the user-initiated Reports can be initiated (Online and/or Offline).
	Validate that the User Interface can access other Modules as required.
	Validate that the Module is stable enough to allow for testing under positive and negative data entry and retrieval Conditions

Figure 72 - Sample Test Case for User Interface Stability

If the Module is not stable, the Test Manager should not allow any Functional Testing to begin that requires this Module.

Navigation

Depending on the Module's size, complexity and platform, testing the navigation could either be very straightforward, or extremely involved. Ideally, a map of the possible navigational paths will have been provided in the Requirements Document or the Design Document. If a map does not exist, the Tester must create a complete map of the relationships and document the flow. This should be reviewed with the Developers before testing begins.

In general, these paths identify how a user can navigate through or between the Screens and Modules. Since it is not possible to predict and test every possible combination of pathway selections a user could make while navigating

> **Authors' note:** It is impossible to know in advance how a user will navigate through a set of UI Screens.

through a UI, it is important for the Tester to have as much information about how the Users intend to use the Module, and to minimally understand their expectations for the basic field to field, Screen to Screen, and Module to Module navigational flow.

There are also common testing Conditions that transcend these differences and that should be included in the tests. Some of these Conditions are:

TEST CASE	CONDITIONS
Validate Navigation	Exercise all: • Menu/Sub-menu options • Functions/Sub-Functions • Navigation Keys • Function Keys • Function Keys with modifiers • Shortcuts
	Validate accurate movement from: • One field to/from the next • One record to/from the next • One Screen to/from to the next • One Module to/from the next • The Module to/from Databases and Files • The Module to/from other Modules
	Using all Screen Controls • Open • Minimize • Maximize • Restore • Close • Remain on top, etc.
	After each type of operation for all relevant Screens: • Add • Change • Delete • Update • Duplicate • Insert • Save • Print • Send • Search • Sort • Undo, etc.
	Validate navigation through the following objects using all of the applicable methods (i.e. horizontal scrolls, vertical scrolls, hot keys, Codes, etc.): • Combo Boxes • Dialogue Boxes • List Boxes • Grids, etc.
	Validate the ability to have multiple Screens open simultaneously, if applicable, while using all possible activities (i.e., toggle, cascade, etc.)
	Validate accurate navigation based on data specific Conditions. For example, if the entry of a specific security type on a Security Add Screen dictates the completion of data on a second Screen, validate the navigation to that second Screen.
	Exercise all paths using keyboard and mouse operations
	Navigate with and without Defect Conditions present
	Navigate using all relevant combinations of: • Platforms • Operating Systems • Browsers

Figure 73 - Sample Test Case for User Interface Navigation

Although there will likely be a number of Issues that result from this level of testing, unless the Issue is serious, the Test Manager will normally allow Functional Testing to begin. Also, fine-tuning of the navigation typically continues throughout the Project's entire life cycle. Frequently it is not until the actual End Users start to use the Application robustly either during one of the testing phases, or in Production that they discover changes that they may want to make to the flow of the Screens, etc.

The remaining Technical Business Tests could be run concurrently with the Functional Test or serially depending on the staffing and the Plan.

Appearance

It is important that the Module's appearance match what was expected and follows the organization's Standards. Applications that follow a common Standard are easier to learn and easier to maintain.

TEST CASE	CONDITIONS
Validate Appearance	Layout
	Design
	Sequence of: • Menu Options • Functions • Fields
	Special Features: • Font Styles/Sizes • Highlighting • Shading • Underlining • Italics • Borders • Colors, etc.

Figure 74 - Sample Test Case for User Interface Appearance

Defects resulting from testing the appearance of the Module are not usually serious enough to delay testing.

Field Edits

Hopefully, through the use of one of the previously noted Documents, all of the information that is required to perform this level of testing will be available. In a Large Project, this Task is very often given to a junior Tester since the rules are generally clearly defined, and the preparation, Execution and validation do not require a great amount of business knowledge.

TEST CASE	CONDITIONS
Validate Field Edits	Field Sizes
	Accurate handling of Required/ Optional fields
	Validation of restricted field contents
	Cross-field Edits
	Test using valid and invalid data
	Specialized field types: • Dates/Times • Currencies • Country Codes • Fractions/Percentages • Zip and Postal Codes • Telephone Numbers • Social Security Numbers, etc.

Figure 75 - Sample Test Case for User Interface Field Edits

Field Sizes

Since it is not feasible, and in many cases not necessary, to test every possible data combination for every field, there are a number of techniques to reduce the Scope while ensuring adequate test coverage. One of the ways to do this is through Boundary Testing.

Boundary Testing is a testing technique where the focus is on the lower and upper value limits for any given field. Historically, the majority of Defects fall within the outer limits of a given range, and focusing the testing on these extremes not only makes the testing effort more manageable by reducing the number of Test Cases, but also provides a high level of certainty of covering the most problematic Conditions. This type of testing, coupled with the functionally oriented tests that typically fall in the mid range of the scale, usually provides sufficient data coverage.

To prepare for this test, the Tester will need to know the Field Edits that have been defined for each field. Once that information is known, the following checklist could be followed to prepare a comprehensive set of boundary Test Procedures:

LIMIT		CONDITION
Upper Limit	(Maximum)	Valid Condition
Upper Limit +1	(Maximum +1)	Invalid Condition
Upper Limit -1	(Maximum -1)	Valid Condition
Lower Limit	(Minimum)	Valid Condition
Lower Limit +1	(Minimum +1)	Valid Condition
Lower Limit –1	(Minimum-1)	Invalid Condition

Figure 76 – Sample Boundary Testing

Therefore, in an example where a Broker Commission field for any trade has an acceptable range of $0 – $10,000, the boundary analysis would look as follows:

INVALID	VALID	VALID	VALID	VALID	INVALID	COMMENTS
	Represents lowest limit	Represents lowest limit +1	Represents highest limit -1	Represents highest limit	Represents highest limit +1	
$-1	$0	$1	$9,999	$10,000	$10,001	Represents six test scenarios covering four valid and two invalid boundary Conditions.

Figure 77 – Example of Boundary Testing Conditions

Some additional things to consider when testing numeric fields are:

- Always test with a 0 and a negative number even if they are not within the range of acceptable values. Very often these entries cause Error Conditions, and therefore should always be included within the suite of Boundary Tests.

- To effectively test the totals fields, a Test Case must be built to test for the maximum value that could possibly be encountered. It will be necessary to do some analysis to determine from a business perspective how large that number could possible get, and subsequently determine the input. Factored into the analysis should be an understanding of the maximum value of the individual field, and a Projection of the maximum number of occurrences within a given time period.

Using the previous example of the Broker Commission Field, if there was a Broker Commission Report that Reported total commissions on a monthly basis, it would be necessary to estimate the maximum number of brokered trades that would occur during a peak month, and multiple that number by the maximum dollar amount allowed for one trade. This result should provide a fairly good estimate of the outer boundary for the totals field.

- In general, use all digits (0-9), and vary the positions for all the tests across the numbers that are used.
- Try a repetitive series of numbers, such as 99999, which could have been used by the Developers during their Unit Testing or to identify the end of a process or a file.

Alpha-numeric fields

Some common Conditions that should be tested for alphanumeric fields are:

TEST CASE	CONDITIONS
Validate Alpha-Numeric Fields	With spaces
	With upper and lower case letters
	With letters, numbers, combination of letters and numbers
	With special characters such as: ~ ` ! @ # $ % & * ( ) + = { } [ ]	\ ; : " " ' ' . , ? / <Del>

Figure 78 – Potential Special Characters

Accurate Handling of Required and Optional Fields

The testing of required and optional fields is typically straightforward, as long as there is accurate and complete Documentation. In general, tests should be prepared to validate that all of the fields defined as required are in fact required by the Application, and that all of the fields defined as optional truly are optional. Some situations to consider when preparing these tests are as follows:

- All Required Fields, Subset of Optional Fields
- All Required Fields, No Optional Fields
- All Required Fields, All Optional Fields

The last two situations are the least likely to be covered by the Business Scenarios since these combinations are not all that common. However, since they can exist, it is necessary to create Test Cases to cover them. The first situation is the more realistic situation from a business perspective, and should be covered exhaustively by the Business Scenarios.

Validation of Restricted Field Contents

Very often there is a finite set of acceptable values for any given field. Tests must be prepared to adequately validate each field using all, or a subset, of these values. Since it is not always feasible, nor necessary, to test for all of the possible situations, the Test Manager must decide which values should be tested.

One of the key things to consider when making this decision is whether or not there are processing implications that result from a particular selection. Minimally, a Test Case should be prepared that uses all of the values that have unique processes associated with them.

- If, for example, there are five different options within the Tax Treatment Code field on a Security Add Screen, and each one causes the process to use a different tax treatment, a Test Case should be created for all of the options.

- However, in the case where there is an option to select one Account Officer Name (which is only an informational field) out of a possible listing of 100 valid names, on an Account Opening Screen, there is no need to test all 100 options.

In this case, it would make sense to conduct some Boundary Testing to ensure that the first and last entries are retrievable, etc., and to test the names with special characters.

Cross-Field Edits

Frequently, entering information into one field triggers a requirement that related information be entered into subsequent fields and/or Screens, or that a specific value or range of values be entered into a related field.

- An example of the first Condition is the entry of a disbursement with instructions to disburse funds via a wire or ACH. In this example, it is likely that the receiving bank's ABA and Bank Name along with an Account Number and Name would be required.

- An example of the second Condition is a rule stating that the Settlement Date of a trade must be greater than or equal to the Trade Date.

Each of these Conditions must be understood, Documented, Executed and validated.

Specialized Field Types

Every organization typically makes decisions about how they want specialized data to be processed and presented by the Application. To write these Test Cases, the tester will need to know the rules that have been defined for each field type and test for both valid and invalid Conditions. A sampling of some of the more common specialized data Test Conditions includes:

TEST CASE	CONDITIONS
Validate Specialized Field Types:	Dates/Times: • All relevant Domestic Formats • All relevant International Formats
	All Relevant Currencies • With decimals or without? • Use of commas or periods to separate numbers? • If with decimals, how many? • With currency signs/abbreviations, or without? • With negative numbers displayed with a negative sign, parentheses, or in a specific color?
	Fractions/Percentages • Negative/Positive • With/Without % • With/Without decimals

	Zip/Country Codes
	• Domestic
	    – 5 digits
	    – 9 digits with hyphen
	    – 9 digits without hyphen
	• International
	    – All unique zip Code structures
	    – All unique country Code structures

	Telephone Numbers • With/without Area Codes • With dashes/without • Domestic/International
	Social Security Numbers, etc • With/without dashes

Figure 79 - Sample Test Case for User Interface Specialized Fields

It is important to note that the rules might change depending on the output medium. For example, to make it easier to read the output, currency signs might be omitted from currency values on all Screens and internal operational Reports, but included on Client Statements.

As with all of the other Conditions, the options governing the handling of these fields must be understood by the Tester, and validated on all input and output mediums.

Defaults

Information can be defaulted in a number of different ways. It is important to know how and when these defaults occur in order to prepare appropriate Test Procedures to cover these Conditions. Some of the more common types of default Conditions are noted below:

TEST CASE	CONDITIONS
Validate Default Data Conditions	Where information that is entered on one Screen gets carried forward to subsequent Screens: • Client Name • Security Identifier • Policy Type, etc.
	Where information gets defaulted in one field as a result of information that was entered in a previous field.
	For example, if a trade is entered for a US equity, the Settlement Date will automatically default to Trade Date + 3, or a Location/ Registration will also default to a specific value based on the security that was entered.
	Where information gets defaulted by some other external Condition, i.e., Transaction Date defaulting to the current business date

Figure 80 - Sample Test Case Conditions for User Interface Defaults

Test Cases should be prepared to validate each default rule that exists. The key to streamlining these tests is to cover as many Conditions as possible in each Test Case. For instance, with a case that tests the purchase of one US security, all three of the above Conditions (and possibly many more) could be covered.

Warning and Error Messages

Test Conditions should be created to test every Warning and Error Message. This is usually called Error Handling Testing. Typically this information is not easy to obtain early in the Project since the Development of meaningful Error Messages does not generally rate a high priority. However, it is important that this testing not be overlooked since these Messages are the user's primary source to understand what they or someone else might have done wrong.

> **Authors' note:** Don't forget to test the Error Messages.

These tests should have two goals:

- The first goal is to ensure that if the Condition requires a Warning Message, the Module displays the Warning Message; and if a hard Error exists that should prevent data from either being entered and/or posted, that the Module actually prevents the entry or posting of the data.
- The second goal of the test is to validate the accuracy and meaningfulness of the Warning/Error Messages themselves.

This testing could be approached from two different perspectives.

- The first (and safest) would be to methodically do a field-by-field evaluation and write a complete suite of Test Procedures that will enter invalid data and/or perform invalid operations.
- In the second approach, the testing team waits until all of the Warning/Error Messages have been fully defined, and works backwards from them to establish the Test Conditions that will invoke them. The Risk with this second approach is that if a Developer overlooked a Defect Condition, it would never be identified and tested. Also, since these Messages typically do not get fully defined until late in the process, it

does not leave a lot of time to do what often turns out to be a significant amount of testing.

The Test Manager should determine the approach to take, and how best to handle the potential shortcomings. If the decision is made to take the first approach and methodically evaluate each field and process, the following list can be used as a guide to determine the general types of Conditions to test:

TEST CASE	CONDITIONS
Validate Error Messages	As a result of the entry of values outside of the acceptable ranges
	As a result of omitting required fields either in a single field, or part of a cross-field edit check
	As a result of an attempt to do an illegal operation, i.e., attempt to delete a record that doesn't exist
	As a result of data entered out of sequence, i.e., 'from' range higher than a 'to' range
	As a result of an attempt to perform an operation when access has not been granted, i.e., an unauthorized user attempting to approve a transaction, etc.
	As a result of an attempt to perform an operation when not all of the required Conditions have been met, i.e., attempt to run a Client Statement when there are missing prices, FX rates, etc.

Figure 81 - Sample Test Case for User Interface Error Messages

Although the focus of this discussion is primarily for a GUI-based Application, the Test Manager/Tester will need to decide how far to take the validation for each GUI test.

- For example, there could be separate tests prepared to validate the way the workstation handles the input of a maximum dollar amount and the validation is only performed by the UI.
- On the other hand, the test could include the validation of how that same scenario is handled by the Reports and Interfaces. In this case, the validation includes the Screens, Reports and Interfaces.

In most cases, the Business Technical Tests will validate the first point and the Functional Test will validate the second.

Security

Test Conditions should be created to verify that only authorized Users can access the Application.

TEST CASE	CONDITIONS
Validate User Layer Authentication	As a single Independent User
	As multiple Independent Users
	As a single User within a User Grouping
	As multiple Users within a User Grouping
	As a single Application User
	As a multiple Application User
	As a valid User
	As an invalid User

Figure 82 - Sample Test Case Conditions for User Interface Security

Reports

Reports are tested in two ways:

- By systematic functionally oriented testing that is based upon the Business Functional Scenarios
- By systematic testing that is based upon the Business Technical Scenarios

As was the case with the User Interface, there are several Common Tests that can be exercised regardless of the type of Report that is being tested. The Business Technical Scenarios for Reports include:

- Completeness
- Stability
- Format
- Field Edits
- Warning and Error Messages
- Report Processes

The testing of the Report's business content is performed during Functional Testing, and the Technical Testing is covered in this section.

Test Preparation

The source of the information required to test Reports could be in any number of Documents and Document types. The two major Document types are typically Technical Documents and/or User Documents. The names of the Documents may vary by organization; however, any one or a combination of these sources will provide enough information for the tester to thoroughly test all of the reports. The Documents most frequently used to test Reports are:

- Functional Requirements
- Data Design Document
- GUI Layout/Navigation/Report Document
- Functional Requirements (legacy system – if applicable)
- User Manuals (legacy system – if applicable)

Although the majority of the Defects in Report formatting are identified during the Business Technical Tests, the realistic nature of the business situations that are covered in the Business Functional Testing often yields additional Conditions that were not considered by the Technical Scenarios.

Completeness

As soon as each Report can be produced there should be a review to validate completeness as noted in the following table:

TEST CASE	CONDITIONS
Validate Completeness of Reports:	Validate that all anticipated Modules have been delivered.
	Validate that all of the Reports within each Module have been delivered

	• All Headings present
	• All Fields present
	• All Pages present
	    – Title/Cover
	    – Totals
	• All Sections present

Figure 83 - Sample Test Case for Report Completeness

Stability

Reports must be tested for stability before the Functional Test can begin. Some of the key questions to ask are:

TEST CASE	CONDITIONS
Validate Report Stability	Validate that all required data is available
	Validate that all of the defined reports are available • Scheduled Reports • Reports on Demand

Figure 84 - Sample Test Case for Report Stability

Format

Testing the Format of a Report includes the following Conditions.

TEST CASE	CONDITIONS
Validate Format	Layout
	Spelling
	Font Sizes
	Font Styles
	Highlighting/Underlining/Italics/Borders/ Colors
	With data/without data
	With Defect Conditions/without Defect Conditions
	With data in every field
	With the minimum and maximum amount of occurrences of data, i.e., rows, columns, etc.
	Placement, Alignment for all: • Column Headings • Headers • Footers • Logos

Figure 85 - Sample Test Case for Report Format

Field Edits

Each Report should be capable of reflecting the data that is required, and the space available for each data element should be large enough to include the largest possible value.

TEST CASE	CONDITIONS
Validate Field Edits	Field Sizes
	Accurate presentation of Required/ Optional fields
	Accurate processing of valid and invalid data and requests
	Specialized field types properly formatted: • Dates/Times • Currencies • Country Codes • Fractions/Percentages • Zip and Postal Codes • Telephone Numbers • Social Security Numbers, etc.

Figure 86 - Sample Test Case for Report Field Edits

Warning and Error Messages

These conditions are the same as for User Interfaces.

Report Processes

The following table outlines the general processes that should be tested for all Reports:

TEST CASE	CONDITIONS
Verify Report Processes	Sequencing of: • Headings/Sub-headings • Data within each heading/sub-heading
	Filtering: • By one variable • By multiple variables • By all variables
	Sorting: • Headings/Sub-headings • Data within each heading/sub-heading
	Totals/Subtotals: • Placement • Alignment • Printing of 0's, minimum amounts, maximum amounts • Printing of negative numbers • Printing of maximum amounts as negative values • Printing of minimum/maximum non-numeric fields • Accuracy of all sub totals/totals • Presence/accuracy of totals when data overflows to multiple pages
	Other Calculations: • Percentages • Ratios
	Page Breaks: • Page breaks at logical points • Accurate page numbering
	Error Conditions: • Overflow data Conditions • Missing/Bad Data

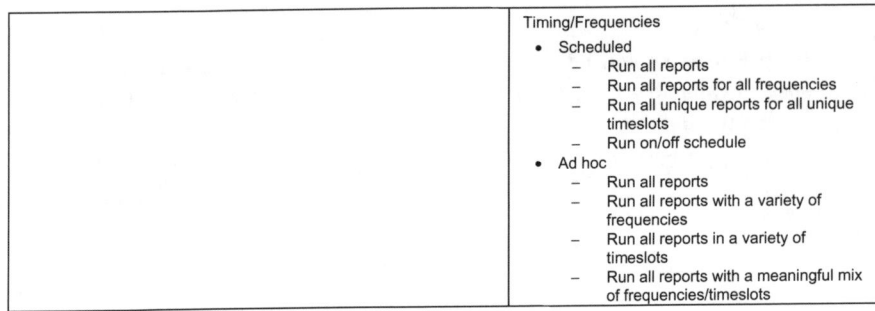

Figure 87 – Sample Test Case for Report Completeness

Once again, this structured testing should be coupled with the Business Functional Scenarios to provide a relatively high coverage level.

When planning for this type of test, it makes sense to re-use as much test data as possible. For example, if the UI Common Tests were run before the Report Common Tests, many of the data combinations that were prepared for the UI tests could probably be re-used to validate similar Conditions on the Reports. While the UI test data likely will not cover all of the Report Conditions it should cover a substantial number of the Conditions, which will save valuable time for the set-up and creation of Test Conditions and test data.

Authors' note: Develop tests that can be reused.

Similarly, these same tests can also be used to validate a portion of the Common Conditions that are needed to test various Files and Interfaces. The following section provides information on the types of Common Tests that should be considered when testing all types of Files and Interfaces.

Files and Databases

There are several common Test Conditions that can be used to test any type of File structure. This group of tests includes the same Conditions that should be included in UI and Report tests, as well as File-specific Conditions that cover the testing of various Record and File Conditions, sequencing and timing, and various transmission Conditions.

As with UIs and Reports, File testing is usually conducted independently from the functionally oriented testing, and paring the two testing processes will result in a high level of coverage. In general, these tests should simulate any Condition that could possibly occur during the creation, transmission, editing, posting, and reporting of data to/from a File. The Technical Scenarios for Files are:

- Completeness
- Stability
- Data Specific
- Record/File Specific
- Warning and Error Messages
- Security

These five Technical Scenarios are covered in the following sections.

Test Preparation

Information about the various files can be found in many of the same Documents that are used for UIs and Reports, including:

- Business Requirements Documents
- Functional Specification Documents
- Data Design Documents

Completeness

The tests for File/Database Completeness should cover:

TEST CASE	CONDITIONS
Validate Completeness of Records/Files	Validate that all anticipated Files/Databases have been delivered.
	Validate that all of the Tables within each File/Database have been delivered
	Validate all record/file structures
	Ability to generate, send, receive and post data from all records/files

Figure 88 - Sample Test Case for File/Database Completeness

Stability

Files and Databases must be tested for stability before the Functional Test can begin. Some of the key questions to ask are:

TEST CASE	CONDITIONS
Validate Files/Databases Stability	Validate that all required Files/Databases are available

	• Validate that each File/Database performed correctly under normal conditions
	    – Single Users
	    – Multiple concurrent Users
	• Validate that each File/Database performed correctly under abnormal conditions
	    – Empty Files/.Databases
	    – System crashes
	    – Invalid data

Figure 89 - Sample Test Case for File/Database Stability

Data Specific

The same type of data permutations that were defined for both the GUIs and Reports should be run through each File type. It cannot be assumed that because a Condition passed a test for either of the previously tested Components that it will automatically pass the File test as the data gets mapped, (usually) translated, and ultimately posted to a Database and/or reported.

In general, the objective of testing these Conditions is to verify that the File's Design can support all of the Business Requirements, and ultimately that the receiving Application(s) can process, store, and Report on each of these Conditions. These tests should include all of the Conditions that are defined in the UI section (as they relate to the specific File being tested) as well as a number of other File specific Conditions.

The Data Specific tests for Records/Files should cover:

TEST CASE	CONDITIONS
Validate Record/File Data	Field Sizes
	Accurate handling of required/optional fields
	Accurate handling of valid and invalid data
	Validation of field contents that are restricted
	Default values
	Cross-field edits
	Accurate handling of specialized field types
	Data transformations
	Relevant data Defect Conditions
	Field Edits

Figure 90 - Sample Test Case for Data Specific Conditions

As previously mentioned, it saves time and effort if the test data that was created for the UI Technical Testing is processed through to the related Files. Although this existing test data will not cover all of the Conditions that are needed to satisfy the universe of the File tests, it will cover enough to make it worthwhile to Plan for the data to be reusable.

Record/File Specific

There are a number of common Record and File Conditions that should be tested during this phase of the test. This should include Conditions where the Record(s)/File(s) should process successfully, as well as Conditions where they should not. The following is a sampling of the Conditions to consider:

The specific tests for Records and Files should cover:

TEST CASE	CONDITIONS
Validate Record/File Specific Conditions	Content • Validate that the Record/File layouts match the Requirements • Validate accurate security encryption • Validate each Record/File type where data is/is not present
	Unique Sequencing/Scheduling Possibilities • File dependencies • Time dependencies • Condition dependencies • Handling of a partially/fully processed File • Processing of a single File/multiple Files • Concurrent Database updates by File/GUI
	Error Handling • Erroneously labelled Records/Files • Invalid file/ record types • Invalid record/file sequence • Duplicate Records/Files • Bad data in one, some, all Records/Files
	Other • Should a File be processed if a subset of the data is bad? If yes, what is the exception processing procedure? • Is there a separate File name for each event/day?
	Header/Trailer
	Empty File

Figure 91 - Sample Test Case for File Specific Conditions

Warning and Error Messages

These conditions are the same as for User Interfaces.

Security

Test Conditions should be created to verify that only authorized Users can access the Files/Databases.

TEST CASE	CONDITIONS
Validate User Access Restrictions	Validate that authorized Users can access each File/Database for inquiry and update
	Validate that unauthorized Users cannot access Files/Databases for inquiry and update

Figure 92 - Sample Test Case Conditions for File/Database Security

Application Interfaces

The Application that is being tested will probably connect to other existing Applications. These Interfaces should be exercised by the functionally-oriented tests, but only after they have successfully passed the Common Tests identified by the Technical Scenarios, which include:

- Completeness
- Stability
- Format
- Data Specific
- Warning and Error Messages
- Security
- Interface Reliability

Each of these Technical Scenarios is discussed in the following sections.

Test Preparation

The Documents most frequently used to test Application Interfaces are:

- Functional Requirements
- Data Design Document
- User Manuals (legacy system – if applicable)
- Technical Documentation for the 'other' Application

Completeness

The tests for Application Interface Completeness are similar to the tests for File Completeness and should cover:

TEST CASE	CONDITIONS
Validate Application Interface Completeness	Validate that all anticipated Modules have been delivered.
	Validate that all of the Application Interfaces within each Module have been delivered

	Content • Validate that the Application Interface layouts match the Requirements • Validate accurate security encryption • Validate accuracy of data on header/trailer records
	Transmission/Processing • All unique holiday processing • Restart/Recovery process
	Unique Sequencing/Scheduling Possibilities • Time dependencies • Condition dependencies • Handling of a partially/fully processed File or Message • Concurrent Database updates
	Other • Is there a separate designator for each event/day process?

Figure 93 - Sample Test Case for Application Interface Completeness

Stability

Each Application Interface should be tested to ensure that the Application reacts correctly under various Conditions, such as:

- If the Interface File or Message is present, or is not present
- If the Interface fails during processing
- If the Database fails during updates

The tests for Interface Reliability should cover:

TEST CASE	CONDITIONS
Validate Application Interface for Stability	Error Handling • Record(s)/File(s)/Message(s) not sent/received • Erroneously labelled Messages/Records/Files • Invalid header date • Invalid file/ record types and sequence • Invalid trailer amounts • Duplicate Messages /Records/Files • Old Records/ Files Transmitted as current • Remaining in the queue while current Record/File attempting to be transmitted • Bad data in one/ some/ all Records/Files • Transmission partially successful • Transmission fully successful
	Other • Should a File be sent if there is no data? • Should a File be processed if a subset of the data is bad? If yes, what is the exception processing procedure?
	Unique Sequencing/Scheduling Possibilities • File dependencies • Time dependencies • Condition dependencies • Handling of a partially/fully processed Messages /File • Processing of a single File/multiple Files • Processing of a single Message/multiple Messages • Concurrent Database updates by File/GUI
	Other • Should a Messages/File be processed if a subset of the data is bad? If yes, what is the exception processing procedure?

Test Case	Conditions
	Processing • Header/Trailer • Empty File/Database • Incomplete Message

Figure 94 - Sample Test Case for Application Interface Reliability

Format/Content

The Application Interfaces could include Messages and Files.

Test Case	Conditions
Validate Record/File Specific Conditions	Content • Validate that the Messages /Record/File layouts match the Requirements • Validate accurate security encryption • Validate each Messages /Record/File type where data is/is not present • Verify that Messages and File transfers can function correctly with valid and invalid data

Figure 95 - Sample Test Case for Application Interface Format

Data Specific

The Data Specific tests could include the following:

Test Case	Conditions
Validate Data Specific Conditions	Field Sizes
	Accurate handling of required/optional fields
	Accurate handling of valid and invalid data
	Validation of field contents that are restricted
	Default values
	Cross-field edits
	Accurate handling of specialized field types
	Data transformations

Figure 96 - Sample Test Case for Data Specific Application Interfaces

Warning and Error Messages

These conditions are the same as for User Interfaces.

Security

Test Conditions should be created to verify that only authorized Applications and Users can access the Application Interface.

Test Case	Conditions
Validate Application Interface Authentication	Validate that authorized Users can access each Interface for import and export
	Validate that unauthorized Users cannot access each Interface for import and export
	Validate that authorized Applications can access each Interface Authentication for import and export
	Validate that unauthorized Applications cannot access each Interface Authentication for import and export

Figure 97 - Sample Test Case Conditions for Application Interface Security

Database/Environment Preparation

In general, Business Technical Tests will not require a dedicated database or any special environmental conditions unless it is necessary to test the User Interface under various specific technical conditions, such as Operating Systems or Browser types.

Test Execution

The process of executing the Business Technical Tests is combined with the execution process for the Business Functional Tests. In some cases the Technical Tests should be completed before the Functional Tests in order to simplify the process, and in other situations, the order of the tests is immaterial.

Issues Tracking

The Issues Tracking process for Business Technical Tests is the same as for Business Functional Tests, and is discussed in Chapter IX.

Summary

Business Technical Tests focus on four primary Application Components:

- User Interfaces
- Reports
- Files and Databases
- Application Interfaces

And, the testing process consists of:

- Test Material Preparation
- Database/Environment Preparation
- Test Execution
- Issues Tracking

The Test Conditions that are contained in this chapter should be viewed as a core group of tests that can be expanded or reduced based on their relevance to the Module that is tested. Although this set of Conditions should be meaningful regardless of the type of Application, platform, operating system, browser, or method of data transfer, there will undoubtedly be Conditions that are either not relevant, or not listed and relevant.

The Test Manager and Testers must modify this list to create a meaningful and workable set of Conditions for their unique circumstances. Ultimately, the combination of these Technical Test Conditions and the Functional Tests, along with the Business Integration, Technical Tests, UAT and Parallel Tests will provide a very high level of test coverage.

XI. MANAGING TEST PROJECT PROCESSES

In general, the Test Manager is responsible for overseeing all of the testing Tasks that fall into four general management categories:

- Planning
- Executing
- Controlling
- Closing

In Chapters II – VIII we covered the Planning activities that include how to determine the types of Test Project Categories, how to staff the test organization, and how to build the Summary Test Plan, Detailed Test Plans and the Test Project Plan. In Chapters IX and X, we reviewed the remaining Planning activities that included how to write the detailed test Documents, and how to prepare the Test Execution Tasks.

In this chapter and the next, we will review the primary Tasks that are involved in the Controlling process. It is within this management process that the Test Manager monitors and measures the actual Test Results, and takes corrective action to keep the Plans on track. Towards that end, there are seven main topics that the Test Manager must continually evaluate:

	TOPIC	SOLUTION
1	Are the testing Tasks being Executed as Planned, and on time?	Task Management
2	Are the Tests being Executed as Planned, and on time?	Test Management
3	Is the quality of the Application meeting the agreed upon Standards?	Quality Management
4	Has there been any Scope Creep?	Scope Management
5	Is the cost within the allocated Budget?	Financial Management
6	Are there pending Risks that could impact the timeliness, quality, cost and usefulness of the product?	Risk Management
7	Will the Application meet the End Users' requirements?	Chapter XIV – Closing the Project

Figure 98 – Test Management Solutions

The first, second and third questions are covered in this chapter, the fourth through sixth questions are covered in the next chapter, and the final question is covered in Chapter XIV.

To help answer these questions, the Test Manager must have access to summary level information that provides the current status of the milestones that have been defined for each of the major deliverables. The Reports that contain this summary level information are typically called MIS since the Reports result from some manual or automated Management Information System (MIS) process.

Authors' note: Spend the time now to define your MIS and you will always know where you are throughout the Project.

Although the creation of this MIS can be time consuming initially, we strongly recommend that the Test Manager invest the necessary time at the beginning of the Project, as it will become an essential tool for him/her to identify potential trouble spots. The earlier in the process that a problem is detected, the more time there is to correct it and make the necessary adjustments to keep the Project on track, and the less that the corrections will cost.

To effectively manage these activities, there are four steps that must be taken:

- Define meaningful MIS
- Assess the Status of the Tasks using the MIS
- Adjust the Plans as required
- Communicate the Status and modifications to the Plan

These four steps are used to explain the various management processes that are discussed in this chapter and the next.

Although the format of the MIS can vary from organization to organization, the need and the content of the information remain fairly constant. The following sections provide sample Reports that have worked for the authors while supporting the Implementation of many large Applications in the Financial Services Industry.

> **Authors' note:** The Summit Group has defined a three-step process that is useful in almost every management situation:
> Measure
> Monitor
> Manage

Task Management

One of the most important areas the Test Manager must focus on includes the monitoring and adjusting of all of the tasks that are established in the Project Plan. These tasks include all of the management and planning tasks as well as all of the tasks associated with the actual Execution and verification of the tests once the testing is underway, through to the tasks that are related to Closing the Project.

The development of the Project Plan was covered in Chapter VIII, and the management tasks are discussed in this chapter and the next.

Test Management

There are six steps involved in managing the tests:

- Test Preparation
- Test the Test Process
- Execute/Control the Tests
- Monitor the Progress of the Tests
- Re-Test as Required
- Close the Project

> **Authors' note:** As with all of the other Documentation noted, this information should be tailored to meet the specific requirements of your own organization.

Test Preparation was covered in Chapter IX, and the Closing activity is covered in Chapter XIV. The remaining activities are described in this chapter.

Test the Test Process

The Test Manager should ensure that the process that they established will work by testing it on a small portion of the Project. The manager should ensure that the Testers are trained, the process flow is efficient and the Documentation is appropriate.

Execute/Control the Tests

This could be a relatively straightforward Task when the Project is small in Scope and complexity; however, the bigger and more complex the Project, the bigger and more complex the management effort becomes. In a Large Project, regardless of the SDLC that has been chosen, there will almost always be overlapping activities. This usually happens for two reasons:

- Detailed Planning for future Test Phases typically takes place concurrently with the Execution of the current Test Phase.
- Large complex Testing Projects inevitably come under timing pressure and some of the Test Phases may be overlapped to stay on schedule.

The following chart is an overview of how the Plans for a complex Project might start out:

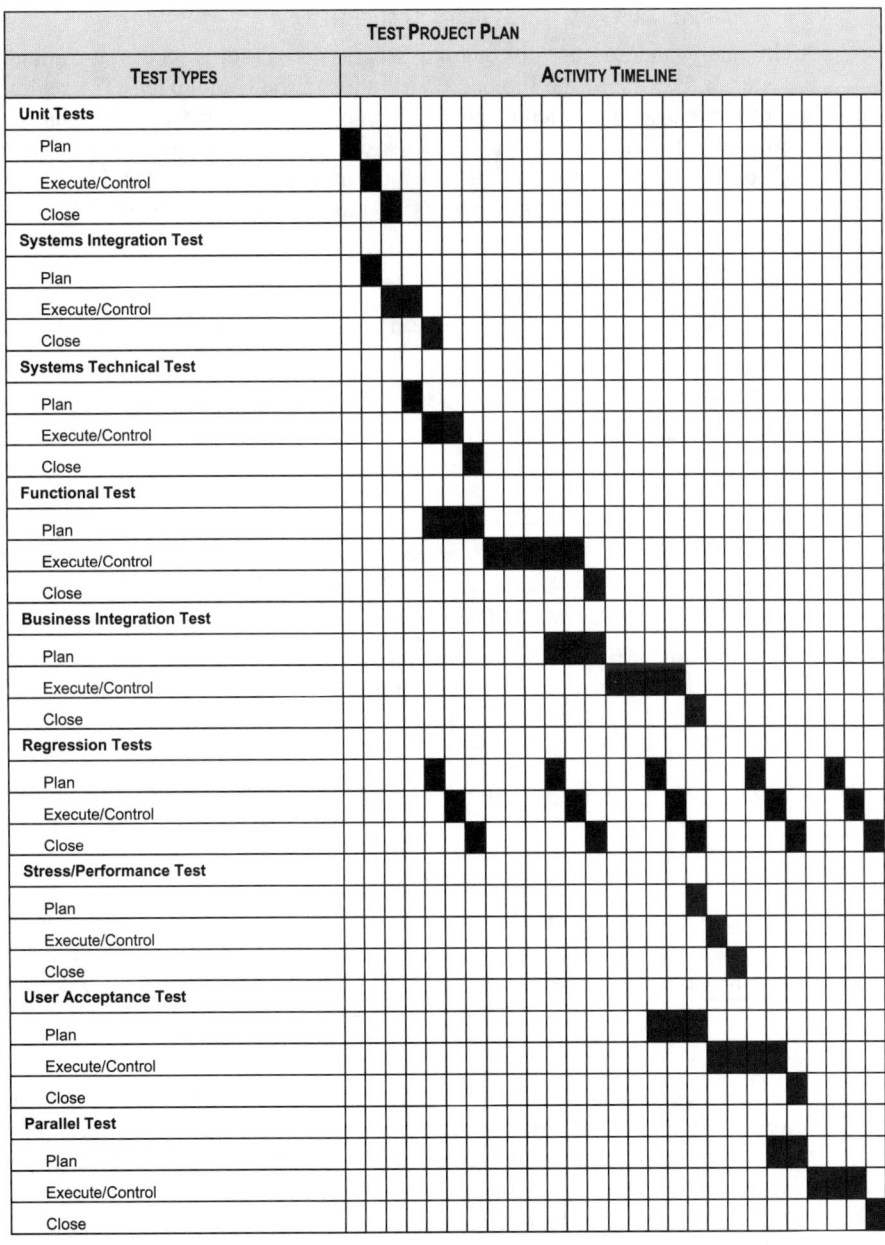

Figure 99 – Sample Timeline for a Complex Project

In looking at this chart, which reflects an almost best case scenario, the Test Manager may have to concurrently oversee up to three major activities. There are overlapping Tasks that occur during the Unit, System Integration and Systems Technical Test Phases that require the simultaneous management of:

- Closing Tasks of the Unit Tests
- Execution/Controlling Tasks of the Systems Integration Test
- Planning Tasks for the Systems Technical Test

Additionally, there is usually ongoing Regression Testing occurring throughout the entire life cycle of the Project, which adds to the complexity. The reality is, as problems arise this picture gets even more complicated and the number of overlapping Tasks can continue to increase in number, thereby creating an even more difficult management effort. As a result, it is crucial for the Test Manager to have accurate and concise information in a usable format to facilitate timely reviews.

Define Meaningful MIS

The Test Manager must focus on the testing Tasks themselves, and whether they are:

- Executed as Planned
- Completed when scheduled
- Documented as defined
- Properly tracking Defects and Defect resolution

Most of the information that the Test Manager needs in order to assess the Status of the testing Project and the testing itself is included in two core Documents: the Project Plan and the Execution Checklist.

- The Project Plan is the central Document that contains a list of all of the Tasks that are required by the testing Project, as well as their Status. This includes the Planning Tasks, Implementation Tasks, Code delivery schedule, and Execution Tasks.
- The Execution Checklist is based upon the Test Calendar and provides detailed information about the Status of the actual tests once testing has begun.

The Test Project Plan was covered in Chapter VIII, and the Execution Checklist was introduced in Chapter VII. The Execution Checklist, and two other Reports that could be used by the Test Manager to monitor the Status of the Project, the High Level Execution Status and the High Level Test Case Status, are described in the next section.

Execution Checklist

The Execution Checklist becomes the Test Manager's primary tool to track every detail of the testing activities once Test Execution has begun. Additionally, the information contained in this Report is often arranged in a number of different ways to provide the Test Manager with various Views of the test Status. The following is a brief description and sample layouts for some of these Reports.

The Execution Checklist example provided in Chapter VII looked as follows:

EXECUTION CHECKLIST

SYSTEM DATE	ACTUAL DATE	TEST PROCEDURE /TASK	TESTER
10/31	9/09	F-3.1.1.1 Buy Security - Equity	Terri Tester
		F-3.1.1.2 Buy Security - Equity	Terri Tester
		F-3.1.1.3 Buy Security - Equity	Terri Tester
		F-9.1.1.1 Settlement	Tom Tester
		F-9.1.1.2 Settlement	Tom Tester
		F-9.1.1.3 Settlement	Tom Tester
		Notify Technology of completion of input for the day	Mike Manager
		Process Pricing Files	IT Department
		Run Trade Batches	IT Department

Figure 100 - Sample Execution Checklist

A number of helpful tracking Reports can be produced from this base information. The first Report could be one where the actual completion date was added to this format to allow detailed tracking as displayed below.

DETAILED EXECUTION CHECKLIST (WITH COMPLETION DATE)

SYSTEM DATE	ACTUAL DATE	TEST PROCEDURE /TASK	TESTER	ACTUAL COMPLETION DATE
10/31	9/09	F-3.1.1.1 Buy Security - Equity	Terri Tester	
		F-3.1.1.2 Buy Security - Equity	Terri Tester	
		F-3.1.1.3 Buy Security - Equity	Terri Tester	
		F-9.1.1.1 Settlement	Tom Tester	
		F-9.1.1.2 Settlement	Tom Tester	
		F-9.1.1.3 Settlement	Tom Tester	
		Notify Technology of completion of input for the day	Mike Manager	
		Process Pricing Files	IT Department	
		Run Trade Batches	IT Department	

Figure 101 - Sample Execution Checklist (With Completion Date)

This Report allows the tracking of every Test Case and every daily process. In some instances, when an automated test tool is used, the actual tracking of the cases that are entered and Executed by the tool can be automatically tracked. However, it is not uncommon for a large number of the Test Cases to be entered manually, even in a large Project. When there is a mix of Execution methods, separate Reports for manual and automated entry should be produced and if time permits, consolidated into one Report.

High Level Execution Status

In addition to tracking the individual Test Cases, the Test Manager might be interested in consolidated Views of the Test Execution and/or Test Case Status.

Information for both of these Reports can be summarized directly from the detailed Execution Checklist. The following is a sample Report format that shows the Execution Status by Execution date.

| FUNCTIONAL TEST ||||||||
| HIGH LEVEL EXECUTION STATUS (AS OF 9/11/02) ||||||||
SYSTEM DATE	TEST CASE	TEST PROCEDURE	PLANNED EXECUTION DATE	ACTUAL EXECUTION DATE	STATUS
10/31	F-3.1 Buy Security - Equity	F-3.1.1.1	9/09	9/09	Complete
	F-3.1 Buy Security - Equity	F-3.1.1.2	9/09	9/09	Complete
	F-3.1 Buy Security - Equity	F-3.1.1.3	9/09		Late
11/01	F-3.2 Buy Security – Corporate Bond	F-3.2.1.1	9/10	9/10	Complete
	F-3.2 Buy Security - Corporate Bond	F-3.2.1.2	9/10		Late
	F-3.2 Buy Security - Corporate Bond	F-3.2.1.3	9/10	9/10	Complete

Figure 102 – Sample Execution Status Report

Frequently, the testing effort cannot move forward because of Issues in one particular area. This Report can show the Test Manager the Test Case Status for each Functional area, and help the Test Manager isolate the problem area to determine the cause and identify possible adjustments.

High Level Test Case Status

This Report allows the Test Manager to measure the Test Case Status for any given Function, and it helps the Test Manager see patterns in how the Test Cases are Planned and how they are progressing compared to the Plan.

For instance, if there is a consistent pattern with many simple Test Cases that are Planned for one day and many complex tests on the next day, the manager could move some transactions from one day to the next to ensure that they all get Executed as Planned.

The following is a sample layout of this Report.

| FUNCTIONAL TEST |||||||
| HIGH LEVEL TEST CASE STATUS |||||||
SYSTEM DATE	PLANNED EXECUTION DATE	TEST CASE	PLANNED # OF TEST PROCEDURES TO EXECUTE	ACTUAL # OF TEST PROCEDURES EXECUTED	STATUS
10/31	9/09	F-3.1 Buy Security - Equity	3	2	Reschedule Test F-3.1.1.3 for 11/02
11/01	9/10	F-3.2 Buy Security - Corporate Bond	3	2	Reschedule Test F-3.2.1.2 for 11/02

Figure 103 – Sample Test Case Tracking Report

Each of the Reports/views presented here is an example of the various types of tracking Documents that have proven useful to Test Managers in our experience. As we mentioned before, the key is to have Documentation and Reports available that balance the Test Manager's need for information with the effort that is required to produce the Reports.

> **Authors' note:** The format is not important, but the content is.

Unless mandated by the organization, the format that a manager selects is purely a personal preference unless it is mandated by the Testing Tool that is selected. The important point is to have some Report that allows the Test Manager to review the Status quickly, and as frequently as needed.

Reviewing the Status

In general, the Test Manager does not need to review the status of every activity with the same frequency or intensity. There are some Tasks that will need to be reviewed daily, while others only need to be reviewed on a less frequent basis. For example, while it is important to review the Status of many of the testing Tasks on a daily or weekly basis, a review of the financials might take place only monthly.

The Test Manager should establish the review-frequency for each of the Reports that meet the Project's objectives; however, a possible schedule could look as follows:

POSSIBLE TASK REVIEW SCHEDULE			
REPORT	DAILY	WEEKLY	MONTHLY
Project Plan			
By Milestone			X
In-Progress Tasks	X		
Late Tasks	X		
Tasks Specified To Be Within A Certain Timeframe		X	
By Resource		X	
Detailed Execution Checklist	X		
High Level Execution Status		X	
High Level Test Case Status		X	

Figure 104 – Sample Report Review Frequency

The Detailed Execution Checklist should be the Document that is used to track testing progress on a day-to-day basis. This Report should be a roadmap for each day's Test Execution activity. This Document gets adjusted frequently as the events of the day unfold and situations occur that require continuous shifts in activity or timing in order to accomplish the Planned tests.

The High Level Execution Status and High Level Test Case Status Reports should be reviewed at least weekly to allow the Test Manager to get a higher level and longer-term impression of how the overall testing is progressing. Once the actual dates start to get filled in, these Reports provide insight into possible trends that might require systemic remediation, and provide an ability to review quickly the Status of key milestones.

Additionally, these Reports can also be used as a valuable source of information to support the day-to-day decision making process. For example, if the Test Manager is considering reducing the number of trades that are entered on one date and carrying over a subset of the trades to the following day, a review of the total Test Cases that are Planned in the High Level Test Case Status Report can help the Test Manager decide whether or not it is feasible, and/or if some additional adjustments need to be made.

If an automated tool is used to track the Project, the various Project Plan Views can be continuously up-to-date, and could be reviewed at any time; however, it is more likely that the Test Manager will still review the Reports with some mixed frequency as depicted in the chart above.

Although these Documents provide a substantial amount of information, Test Managers frequently create supplemental Reports that offer different Views of the Test Project's progress to meet their specific needs.

Adjust The Plans

Once testing gets underway, problems will likely be found in every process that is tested. The following is a listing of possible problem areas that might cause testing to get off track, and a Matrix that maps possible solutions for each category. Although there is very rarely a single solution and/or one static set of adjustments that will fix a given problem, there are adjustments, either by themselves or in combination, that have been used successfully to correct the problems most frequently encountered.

Potential Problem Areas and Solutions

The following tables contain some of the more common problem areas that can potentially cause a slippage, along with some recommended solutions. These tables build on the information that was provided in Chapter VI to help manage the Test Calendar, and are designed to assist the Test Manager resolve problems throughout the project.

All of these problems and potential solutions evolve throughout the Project. Some of the more workable combinations are noted in the following series of charts but this is not meant to limit the solutions. As stated previously, the Test Manager needs to be creative in identifying the combinations that will work for specific Issues.

SOLUTIONS		PEOPLE				
		ISSUES				
		INSUFFICIENT STAFFING	INEXPERIENCED TESTING STAFF	STAFFING TURNOVERS	TESTING STAFF UNFAMILIAR WITH THE BUSINESS	UNCOOPERATIVE TEAM PLAYERS
People	Add / Replace People	X		X	X	X
	Training		X		X	
	Overtime	X				
	Test Multiple Shifts	X				
	Outsource	X	X	X		
	Incentives	X		X		
	Appreciation	X		X		
	Add Business People	X	X	X	X	
Scope	Reduce Test Scope	X				
	Reduce Functionality	X				
Hardware and Software	Add Hardware / Test Environments					
	Performance Tweak					
	Quality Review					
	Adjust Job Streams					
	Adjust Job Priorities					
Test Method and Calendar	Ensure Unit Test		X			
	Test in Parallel Mode					
	Phase in Software	X				
	Reduce Number of Test Cycles / Days	X				
	Begin Test without formalized Documents					
	Begin Test with Stubs					
	Collapse Test Phases	X				
	Combine Test Cases	X				
	Increase Test Automation	X	X	X		
Management	Ensure Senior Management Support					X
	Ensure All Other Management Support					X
	Ensure All Other Tools are in Place for Effective Test Management					
	Project and/or Test Managers to Evaluate Impact					

Figure 105 – People Issues and Potential Solutions

SOLUTIONS		HARDWARE / SOFTWARE							
		ISSUES					SYSTEM ISSUES		
		LATE CODE	TEST ENVIRONMENT NOT READY ON TIME	SCOPE CREEP	POOR QUALITY CODE	SLOW TURNAROUND ON DEFECT FIXES	RESPONSE TIME	AVAILABILITY	SIGNIFICANT DESIGN FLAWS
People	Add / Replace People	X	X	X	X	X			
	Training				X				X
	Overtime	X	X	X	X	X		X	
	Test Multiple Shifts	X	X	X	X	X		X	
	Outsource	X							
	Incentives	X				X			
	Appreciation	X				X			
	Add Business People	X							
Scope	Reduce Test Scope	X	X						
	Reduce Functionality	X	X						
Hardware and Software	Add Hardware / Environment Capacity		X				X	X	
	Performance Tweak						X		
	Quality Review				X				X
	Adjust Job Streams						X		
	Adjust Job Priorities		X				X		
Test Method and Calendar	Ensure Unit Test	X	X	X	X	X			
	Test in Parallel Mode	X	X	X					
	Phase in Software	X		X	X				
	Reduce Number of Test Cycles / Days	X	X	X	X				
	Begin Test without formalized Documents			X					
	Begin Test with Stubs			X					
	Collapse Test Phases	X	X	X					
	Combine Test Cases	X	X	X					
	Increase Test Automation	X	X	X					
Management	Ensure Senior Management Support	X	X	X	X	X	X	X	X
	Ensure All Other Management Support								
	Ensure All Other Tools are in Place for Effective Test Management			X	X	X	X	X	X
	Project and/or Test Managers to Evaluate Impact								

Figure 106 – Hardware/Software Issues and Potential Solutions

Solutions		Issues				
		Test Methodology/Status		Management Issues		
		Test Material Not Ready on Time	Inaccurate Test Estimates	Inadequate Senior Management Support	Disinterested Managers of Involved / Impacted Groups	Poor Test Management
People	Add / Replace People	X	X			
	Training		X			
	Overtime	X	X			
	Test Multiple Shifts	X	X			
	Outsource	X	X			
	Incentives	X				
	Appreciation					
	Add Business People	X	X			
Scope	Reduce Test Scope					
	Reduce Functionality					
Hardware and Software	Add Hardware / Environment Capacity					
	Performance Tweak					
	Quality Review					
	Adjust Job Streams					
	Adjust Job Priorities					
Test Method and Calendar	Ensure Unit Test	X	X			
	Test in Parallel Mode	X	X			
	Phase in Software	X	X			
	Reduce Number of Test Cycles / Days	X	X			
	Begin Test without formalized Documents	X	X			
	Begin Test with Stubs		X			
	Collapse Test Phases	X	X			
	Combine Test Cases	X	X			
	Increase Test Automation	X	X			
Management	Ensure Senior Management Support	X	X		X	X
	Ensure All Other Management Support	X	X		X	
	Ensure All Other Tools are in Place for Effective Test Management	X	X		X	X
	Project and/or Test Managers to Evaluate Impact	X	X	X	X	X

Figure 107 – Test Methodology / Status Management Issues and Potential Solutions

		OTHER			
SOLUTIONS		ISSUES			
		BUDGET	AUDIT	POLITICS	MISCELLANEOUS BUSINESS / PROJECT ISSUES
People	Add / Replace People				
	Training				
	Overtime				
	Test Multiple Shifts				
	Outsource				
	Incentives				
	Appreciation				
	Add Business People				
Scope	Reduce Test Scope	X			
	Reduce Functionality	X			
Hardware and Software	Add Hardware / Environment Capacity				
	Performance Tweak				
	Quality Review				
	Adjust Job Streams				
	Adjust Job Priorities				
Test Method and Calendar	Ensure Unit Test				
	Test in Parallel Mode				
	Phase in Software				
	Reduce Number of Test Cycles / Days				
	Begin Test without formalized Documents				
	Begin Test with Stubs				
	Collapse Test Phases				
	Combine Test Cases				
	Increase Test Automation				
Management	Ensure Senior Management Support	X	X	X	X
	Ensure All Other Management Support	X	X	X	X
	Ensure All Other Tools are in Place for Effective Test Management	X	X	X	X
	Project and/or Test Managers to Evaluate Impact	X	X	X	X

Figure 108 – Other Issues and Potential Solutions

Quality Management

The Test Manager must have a way to determine whether the Code will meet the Documented Functional Requirements and ultimately whether it can be effectively used in Production, while ensuring that the testing Project proceeds in a timely manner. A complete Quality Program includes a combination of the pre-Coding reviews of the Requirements, Design and testing materials, along with an on-going review of the Actual Results of the tests. [38]

---

[38] Quality Assurance was discussed in more detail in Chapter I.

There are four major areas of Quality Assurance:
- Pre-Coding Reviews
- Delivery Reviews
- Code Testing Phases
- Management Assessments

The information provided in this section assumes that the:
- Pre-Coding Reviews (Business Review, Technical Design Review and Data Design Review) have been completed
- Management Assessments that were discussed in the previous section of this chapter continue to be performed

This section focuses on the Delivery Review and the evaluation of the Code through testing. The quality of the Code is evaluated primarily by assessing the results of the Execution of the Test Transactions. In examining these Test Results, the Test Manager should be concerned with both the timely completion of the tests, as well as the success rate. This success rate is determined by comparing the Expected Results to the Actual Results and counting the tests that met expectations and dividing by the total number of tests.

Success Rate = Tests Passing / Total Tests Performed

Testing Timeliness = Tests Executed on Time / Total Tests Performed

Deviations from either the Planned outcomes or the findings from any ad hoc reviews should be logged, evaluated, and subsequently acted upon. The Test Manager must constantly monitor the number of tests being performed vs. the Plan and the success rate. If the success rate is lower than Planned, there will be more re-testing than was Planned and the Test Calendar could be affected. To accomplish this, the Test Manager must use the Task Monitoring Reports that were covered in the previous section, as well as some additional ones. These Reports should provide thorough and timely information about the quality and efficacy of the Code being tested.

As with all other Test Management Reports, if these Reports are not already defined by the organization, the Test Manager should evaluate his/her needs, and define the Reports accordingly. Although the format and content of these Reports will vary from organization to organization, there are some key areas that should be considered to ensure a successful Implementation, such as:

Define Meaningful MIS

The MIS should report on the quality of the Code when it is first delivered and on the subsequent quality of the testing. Once testing commences, the Project Manager needs an on-going assessment of the results of the Execution of the full suite of Test Cases.

Code Delivery

Code Delivery verification should occur immediately after the Code is delivered. Its purpose is to identify major disconnects between what the Test Manager thought was being delivered and what actually was delivered. This initial validation can ultimately save the Test Team a significant amount of time if bad Code was delivered, or the wrong Code was delivered. The following list is a guide for this type of validation.

- Can all of the basic processes be performed?
    - On-line
    - Batch
    - Reports
    - Interfaces
- Graphical User Interfaces
    - Have all Screens been delivered?
    - Are all Data Elements present?
    - Is the navigation flow what it should be?
    - Are all of the basic Functions working:
        - Updates?
        - Inquiries?
- Reports
    - Can the Reports be initiated and run?
    - Are the Reports legible when displayed/printed?
- Other Interfaces
    - Can the Interfaces be initiated and run?
    - Does data get from point A to point B successfully?

The Test Manager can use this simple checklist as a Control Report to assess, at a high level, the completeness of the delivery and the stability of the new Code. The content of this list will obviously be unique for each Code delivery, but it can be as simple and straightforward as this example.

In addition to the individual lists for each delivery, it is also important for the Test Manager to keep a cumulative record of the quality of all of the deliveries. The Test Manager can use this information to see trends in the quality of the Code that might require procedural, and/or Test Calendar changes. And, if the ultimate Implementation date needs to slip partially because of bad or late deliveries, this information will be needed to substantiate the role of the Code deliveries in the delay.

The following is an example of a Report that would provide this high level information:

HIGH LEVEL CODE DELIVERY TRACKING					
CODE DELIVERY DATE	FUNCTION	ON TIME?	CODE ACCEPTED?	COMMENTS
9/06	F-3.0 Buy Security	Y	Y	Code passed the initial inspection.
	F-4.0 Sell Security	Y	Y	Code passed the initial inspection.
	F-5.0 Cash	N	Y	Code passed the initial inspection.
	TR-1.0 Daily Activity Report	Y	N	Problem with hard copy Report generation. Decision was made to move forward with the testing because all critical Reports could be Viewed online. Hard copy Reports expected to be working by 9/9. This delay should not affect the overall Test Calendar.
9/19	F-6.0 Receive	Y	N	Entire Code delivery was rejected. Test Team was unable to access this function successfully. If not delivered successfully by 9/20, the end date could be in jeopardy.
	F-7.0 Delivery	N/A	N/A	

Figure 109 – Sample Code Delivery Tracking Report

After it has been determined that the Code is stable and is ready for comprehensive testing, the Test Manager must have information about the quality of the Code once testing is underway. To obtain this information, the Test Manager should:

- Define a series of Reports
- Identify the sources of data for the Reports
- Define a process to prepare the Reports
- Define the Report frequency
- Identify who is responsible to prepare the Report
- Establish a distribution schedule

These activities should be performed before the testing begins so that information is available from the first day of testing. The key is to ensure that when the information repository is built, all of the data fields can be accessed, and can be reported in any format, with any Sort or View of the information.

All MIS tends to evolve over time to support the Test Manager's reporting needs, and any reporting requirements that might be requested on a periodic basis. For instance, to meet an Audit request, the Test Manager may be requested to produce a Report reflecting the Issues that were logged for one particular area of interest, for a specified timeframe. This ad hoc Report may or may not become part of the regularly produced Report suite.

Report Categories

Generally three types of Reports exist within the framework of this MIS: detailed, summary and ad hoc.

- The detailed test Reports are typically the basis for comprehensive Issue reviews by the Testers, Developers, and the Project Management Team.
- The summary level test Reports are primarily used by the Test Manager for Monitoring, Measuring and Managing, and are used as a source of information for senior level reporting.
- Any authority may require ad hoc Reports at any time.

The following is a description of each type of Report, some Sort options, and their intended usage.

Detailed Test Reports

The consistent point about these Reports, regardless of who is using them, is that they typically contain all of the details that are required for an in-depth analysis of the Issue. Depending on who is using the Report, the Sorts may vary to highlight or prioritize the information according to need.

- The Test Team, for example, might produce a Report on a daily basis, listing each new Issue Sorted by date in order to review each Issue at a detailed level and assign priorities.
- The Development Team on the other hand might also get a Report daily, but may Sort it by priority within Functional area to allow the Development Manager to assign the appropriate Developer to each Issue.
- A third Sort might be used by the Project Management Team to review the details of each Critical and High Issue and their proposed solutions to determine if the solution is cost justified.

Each of these Reports will contain similar information, but the Sorts and intended usages are quite different. Another Report that might be useful during this review process is a Report Sorted by Severity. This Report could give the Test Manager a sense of how many open items exist within each Severity category.

As the new Issues are reviewed, this information might be helpful in reprioritizing some of the existing Issues, or prioritizing differently some of the newer items as a result of that information. The detailed Reports also allow the Test Manager to periodically review all of the Issues to identify duplicates, or to review the quality and completeness of the information that is provided by the Test Team.

These are just a few examples of Reports that contain the full details of the Issues. As usual, the Test Manager must decide which of these Reports, or any others, might be helpful.

Summary Test Reports

Summary Test Reports each provides a different View of the testing situation to help the Test Manager focus on existing or potential problem areas. Some of the more frequently used Report Summaries are:

Issue Type

The Test Manager needs to be aware of the mix of the Issues that are being uncovered. A summary Report that shows totals by the type of Issue could help the manager focus on certain types of Issues for further review. Although the ownership for a particular type of Issue might be outside of the domain of the Test Manager, it is often during the testing process that this information surfaces. As a result, it is critical to the success of the entire Project that the Test Manager is aware of all Issue types.

If, for example, a Report indicated that there was a high concentration of Issues caused by Design problems, that information could be helpful to the Test Manager, the Development Manager and the Project Manager. Since Design Issues that are found during testing typically require more repair and testing time than other types, this information could indicate that there may be a need for serious adjustments to keep the Project on track.

Function

A review of the Issues Sorted and summed by Function (or Transaction Type) can identify the most troublesome Functional areas. If there is a heavy concentration of Defects in one particular area, it might indicate a need to do one or more of the following actions:

- Review the Functional Requirements to ensure that they were written clearly and completely
- Review the Design again
- Review the Test Cases to ensure that they are testing the right things
- Review the skills of the Developers who are working on this Functional area

Again, this Report is another tool to help highlight problem areas as early as possible to allow the most amount of time for remediation.

Priority

A Report by Priority or Severity indicates the overall quality of the Code. A high concentration of Critical and High items could indicate a serious situation. There should be a reverse correlation between the lowest number of Issues to the highest Severity, etc. If that's not the case, further investigation is required.

The expected distribution of Issues would normally be something similar to the following chart.

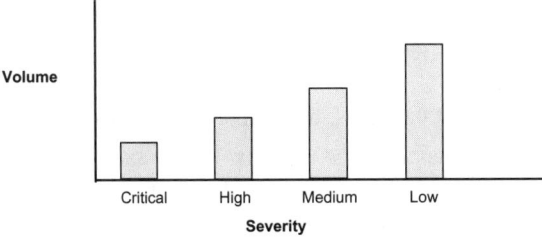

Figure 110 - Expected Distribution of Severity Levels

Some follow-up questions could ascertain whether or not the Priority/Severity is being applied appropriately and by the right people. If it is, and there truly are more Critical/High Issues than all others, is it:

- In a particular Function?
- Across Functions?
- Generated by a particular Tester more often than not?
- Coded by a particular Developer more often than not?

The answers to these questions will help the Test Manager and the Project Management Team determine the appropriate next steps.

Developer

A Sort by Developer is a Report the Development Manager can use along with the Test Manager to identify any Developers who deliver poor quality Code. If this situation exists, it is up to the Test Manager to identify the situation and provide information to the Development Manager (and possibly the Project Manager) to initiate the appropriate remediation.

Tester

It is also important to see if a particular Tester generates an abnormally high or low number of Issues. If either case exists, it could be an indication that the Tester is doing an exceptionally thorough job which would be a good thing, or it could indicate that the Tester is not familiar with the business or Applications, and/or how to successfully conduct and evaluate a test. In any case, this information helps the Test Manager make required adjustments.

Target Completion Date

A Report by Target Completion Date shows how many Defect repairs are due for delivery during a given timeframe. As the testing progresses, the Test Manager will get a feel for how many Defect repairs, etc., can be handled routinely in a given timeframe.

If this Report indicates that there are a larger number of items to be delivered than can typically be accommodated, this information might be the driver for adjusting the timing of the repairs, and/or adjusting the resources needed to test those Issues.

Status

This is another useful Report that provides an indication of the mix of Statuses. If, for example, there is a growing number of items in Re-test that are not getting tested, this could be an indication to the Test Manager that the Testers are not paying attention to items that are returned by the Developers, or it could also be an indication that there are other more difficult problems with the items that remain in a Re-test Status.

An example of this could be that there is a need to wait for the Test Calendar date to move before the testing can be performed, or possibly worse, that it is requires a significant amount of work to recreate the Environment that is needed for a proper Re-test.

Similarly, if there are a high number of deferred items, this information might lead the Test Manager to perform a full review of all of the deferred items to determine if their cumulative effect will have a negative impact on the business. If management decides that not having these fixes completed is a business Risk, adjustments would have to be made to fit them in to the Test Calendar, and possibly revise the Implementation date.

Past Due Reports

There are several variations of Past Due Reports that are helpful to the Test Manager. Some of the important missed deliverables are:

- Missed Target Analysis Dates
- Missed Target Code Delivery Dates
- Missed Target Re-test Dates
- Missed Target Move to Production Dates

A large number of items in any one of these categories can represent a problem for the Test Manager and to the Project. A discussion with the Development Manager would be warranted regarding any slipped analysis and Code delivery dates, the Test Manager should review the missed Re-test dates, and there will be a need for a joint analysis on the missed move dates.

These Reports will enable the Test Manager to see slippages, and provide enough information to identify possible solutions. Some questions to consider are:

- Is there a procedure for these turnovers that is not working?
- Is enough time being allotted for them?
- Does the process need additional management focus?

Totals

Totals Reports are the types of Reports that provide the Test Manager with the highest-level overview of the Status of all Issues. The Project Manager should be able to generate a Totals Report for any Data Element or combination of elements in the repository. Some of the Sorts frequently used are:

- Totals by Issue Type
- Totals by Application
- Totals by Function
- Totals by Severity
- Totals by Status

> **Authors' note:** As mentioned earlier, the key with any testing MIS is to have the ability to generate the type and frequency of Report that will aid in highlighting problem areas as early as possible.

Depending on the flexibility of the reporting process, these Reports could be a collapsed View of the previously defined Reports. If that capability does not exist, they should be created as separate Reports. These Reports are typically used to quickly highlight the most serious trouble spots.

Ad hoc Reports

In addition to the structured detailed and summary Reports, managers will inevitably be asked to produce some ad hoc Reports by senior managers, or the Test Manager will need some information to solve a specific problem. To create these Reports with the least effort, the manager should design a Database of Report information that supports the structured Reports and which also can be used for ad hoc purposes.

The MIS identified in this section is a just sample of the types of Reports that can be needed to support a full quality assessment. Each Test Manager should conduct an independent evaluation of the anticipated reporting needs and ensure that the Reports that are essential to meeting those requirements can be produced.

Report Review Frequency

As with the Task MIS Reports, not all of the Quality Reports need to be reviewed with the same frequency. The review schedule is up to the Test Manager; however, a possible schedule could look like the following:

QUALITY MIS REVIEW SCHEDULE					
REPORT TYPE	REPORT	DAILY	WEEKLY	MONTHLY	AT CODE DELIVERY
New Code Delivery	New Code Delivery			X	X
Detail Level	New Issues By Date	X			
	New Issues By Severity	X			
	All Issues (Any Sort)			X	
Summary Level	Issue Type			X	
	Application/Function		X		
	Priority		X		
	Developer			X	
	Tester			X	
	Target Completion Dates		X		
	Status		X		
	Past Due Reports		X		
	Totals Reports	X [39]			

Figure 111 - Sample Quality MIS Report Review Schedule

The Test Manager should define a core set of Reports at the beginning of the Project and identify a frequency that is appropriate for the Project and the organization. Very often however, the Test Manager will need to run some Reports more frequently, while others are required on an as-needed basis. Because of this, it is very important that the Test Manager have the flexibility to modify the frequency of the Report production.

Adjust the Plans

Quality problems, if not corrected, jeopardize the timeliness and usefulness of the Application that is ultimately delivered. In general, there are Issues that arise from deviations from the test Plans and those that arise from the less formal

---

[39] As the pressure around testing increases, the Test Manager often asks for this Report to be produced on a daily basis.

reviews mentioned earlier. Most of the Issues that are logged will be real Issues while some, when investigated further, will not be seen as a valid Issue.

The following sections review some of the causes for these valid and invalid Issues, as well as recommended solutions for each.

Invalid Issues

There will always be instances where Issues get logged that, after investigation, turn out not to have been a problem. Those situations tend to be resolved without management interference. However, if the Test Manager sees a pattern where Issues are consistently being logged incorrectly, he/she needs to understand the reasons why, and correct the situation.

Hopefully, the Issues Tracking Report will have been designed to capture this type of information. If the Issue categories previously recommended are used, this information would have been included in the Non-Issue category in the Issues Tracking Report.

The chart that follows presents some of the more common reasons for posting Non-Issue items, along with possible solutions.

CAUSE	POSSIBLE SOLUTIONS
Inexperienced Testers	Outsource to a professional testing organization
	Train in testing techniques
	Partner with more experienced Testers
	Establish an Issue review committee to review all Issues
	Replace with more experienced Testers
Testers not familiar with the business	Train in the fundamentals of the business
	Partner with business line resources
	Have business line and/or review committee review all Issues
	Have the Testers focus on more mechanical aspects of testing, i.e., UI edit and navigation, minimums/maximums, etc.
	Replace with people more experienced in the business

Figure 112 – Reason for Non-Issues and Possible Solutions

Valid Issues

Valid Issues and Defects can surface during any Testing Phase. During the Planning process, time should be built in to the Test Calendar to allow for the analysis of the Issues, and the Coding and testing of the repairs. When the actual time for Defect identification, analysis, repair and Re-testing exceeds the allotted time, the Project can be delayed. The Quality MIS that is produced should be used to identify the problem areas, and to take action to correct as many of the root causes as possible.

> **Authors' note:** Leave time in the Plan for Defect repair and Re-testing (and remember that some percentage of the repairs will also fail).

One problem that occurs in many Projects is the debate over the accurate classification of Issues; are they Coding Errors, or do they result from bad Requirements? The Test Manager is often in the middle of this debate as the politics surface and the need to assign blame appears.

In general, when real Defects are identified, they result from one (or more) of the following three causes:

- Bad Requirements (Incomplete, inaccurate or poorly written)
- Bad Design (Inefficient, incorrect or not based upon the Requirements)
- Bad Code (Non-Functioning or not based upon the Requirements)

> **Authors' note:** Finding the balance between analysis and resolution requires an experienced Project Manager.

Regardless of where the fault lies, the Issues need to be resolved in a timely manner. The Project Management Team, including the Test Manager, needs to find a balance between understanding the root cause, and fixing the problem. Too little time in analysis may result in missed opportunities to avoid problems and too much time on analysis leaves less time to resolve the problems.

Understanding the root cause becomes critical when there is a systemic Issue such as bad Requirements. If the Test Manager suspects a problem with the source Documentation, he/she should organize a review session to validate the concerns or dismiss them. This could be time consuming, but:

- If a serious problem does exist, it is better to find out sooner rather than later in order to evaluate the ramifications, and communicate the implications to all appropriate parties.
- If there is no problem, that information also has to be disseminated to all of the Project team members so that they do not continue to mistrust the written Requirements.

Resolving any Issue, regardless of the underlying reason, typically has at least four Components:

- Gain an understanding of the Issue in order to identify adjustments that could be made to resolve the Issue
- Make the necessary adjustments to bring the Test Calendar back in line, when necessary
- Make adjustments to prevent a re-occurrence of the Issue, if appropriate
- Inform the appropriate people about the source of the problem and the adjustments

Some of the more difficult and costly Issues, along with possible solutions to manage them are shown in the following sections. Usually, this information would have been highlighted in one of the MIS Reports reviewed earlier.

Serious Design Flaws – Report by Issue Type

One of the most costly problems occurs when there is a significant Design flaw that requires potential Database and/or Coding Changes. Depending on the complexity of the correction, an Issue like this can seriously delay testing. The Test Manager needs to work with the entire Project Management Team to determine the best course of action.

Design flaws may require that testing be stopped, or the Test Manager may be able to reallocate resources and continue testing in some areas.

Stop Testing

If the worst case occurs and all testing must stop, the Test Manager should review every aspect of the testing effort to see what the Testers can do during the downtime. Some of these steps could be:

- Catch up on test Documentation
- Review the Test Cases to see if they can be made more efficient
- Identify ways to save time once testing resumes, as discussed in the next section

Reallocate Resources

In the more likely event where there is other testing that can go on, there are a number of actions that could be considered:

- Re-deploy people who would have been testing the area that was impacted in order to work on other testing that can move forward. This could possibly complete that testing sooner and free up some people who can work on the revised Code when it is delivered.
- Look to do any one of the normal time saving tactics:
  - Collapse the Test Cases
  - Reduce number of Test Days
  - Test with fewer data Conditions
  - Reduce the test Scope
  - Reduce Functional Scope, etc.
- If the Budget allows, add more Testers and/or involve more Users.
- If there is not enough time, and/or money to get the complete solution in place to make the delivery date, the Project Management Team should determine if it is appropriate to install a partial solution. This decision should be made jointly by the members of the Project Management Team, and possibly by the business' senior management. The business managers must determine if they could work with this temporary solution. Often a partial solution along with some manual workarounds allows the business to be conducted for a short period of time without introducing unnecessary Risk.
- If there are no combinations of adjustments that exist to keep the Test Calendar on track, a decision must be made to extend the Implementation date. This should always be the last resort, but in some instances, there is no other choice. If that is the case, the Test Manager should re-evaluate the Test Calendar, make the necessary adjustments, and communicate the reasons for the change.

Large Numbers of Defects – Totals Report

Hopefully, the quality of the Code, and the ongoing review of the MIS and subsequent adjustments will prevent the number of Defects from reaching the point where they cause the end date to slip. The Test Manager should constantly review these Reports to determine whether or not there has been enough time allocated in the Test Calendar to correct all of the Defects.

To determine this, the Development Manager and the Test Manager must continuously assess how much time will be needed for Defect repair and Re-testing. This is one of the most difficult jobs for the Test Manager to

Control, but without doing this diligently there will never be enough information to make informed decisions.

If it is determined that there is not enough time to repair all of the Defects, the Test Manager needs to assess the possible adjustments, such as:

- Reprioritize and agree to focus on repairing only the Critical and High Severity Defects.
  - This technique is similar to going live with a partial solution to a problem. The business needs to agree that they could move into Production with the cumulative impact of not having the lower priority Defects fixed for some period of time. By making this adjustment, the Development Team has more time to devote to repairing the most important Defects identified by business.
- Re-prioritize some of the Defects originally defined as Critical or High to Medium or Low Severity.
  - This must happen with the End Users' approval, and typically only if there is a commitment to get these Issues fixed, tested and moved into Production very shortly after Implementation.
- Some other time saving tactics are:
  - Collapse the Test Cases
  - Reduce the number of Test Days
  - Test with fewer Transaction Types
  - Reduce the Test Scope
  - Reduce the Functional Scope, etc.

More Critical/ High than Medium/ Low Defects – Priority Report

The existence of this situation could indicate a serious situation. Once again, the Test Manager will need to determine whether it is likely that the Development Team will be able to get all of the Issues resolved and the Test Team will get all of the Issues Re-tested and stay on schedule.

- It is possible that Critical items are less complex than the lower priority items, and that both the Development and Test Managers jointly agree that the repairs and Re-testing can be accomplished within the allotted time. It is important that the Development and Testing Teams make accurate and independent time assessments.
- It is possible that the same item might be complex to correct from the Developer's point of View, but simple from the Testers' perspective. Unfortunately, since this process is not an exact science it requires constant evaluations and creative adjustments to stay on schedule.

High Number of Defects from One Coder – Report by Developer

If there are a large number of Defects that are related to the Code Developed by a particular Developer, the Test Manager should provide that information to the Development Manager. Some possible reasons are: inexperience, lack of business knowledge, misinterpreting the Requirements, bad Requirements, or poor skills.

Each possibility should be explored and the appropriate steps taken.

- If the Project can afford the time, it is possible that either an overview of the Project and/or some additional Training could help if it was an issue of inexperience, not knowing the business, and/or misinterpreting the Requirements.
- If it was bad Requirements, this can also be corrected.
- If due to inadequate skills, it might be better to accept the lost time and replace the Developer.

> **Authors' note:** Developers with poor software development skills do not suddenly improve, nor do Testers with poor testing skills.

Large Numbers of Past Due Deliverables – Past Due Report

As with the three previous situations, large numbers may or may not be a problem. Since there is typically very little excess time built into the Test Calendar, the Test Manager must deal with any slippages. Each missed date should be evaluated independently to determine if the slippage can be absorbed, or if the Plan should be modified.

> **Authors' note:** The Test Manager must constantly identify the problem, assess the damage caused by the problem, find and Implement a solution to prevent the problem from happening again, and determine how to make up for lost time.

Typically the cumulative effect of slippages, occurring across multiple areas, makes the situation more difficult to correct. There is frequently a trickle-down effect with missed dates. Missed 'target analysis dates' often result in missed 'Code delivery dates', which can be followed by 'missed Test Days' and potentially result in 'missed move to Production dates.'

The Test Manager needs to manage all of these dates and understand their long-term impact on the Project. Here again, the key is to understand the cause of the slippage, and determine the best solution to correct it.

Summary

This chapter covered the following key points:

- There are four key Test Management activities including:
  - Planning
  - Executing
  - Controlling
  - Closing
- Within the Controlling Function there are six major activities that the test manager must be concerned with:
  - Controlling the Testing Tasks
  - Assessing the Quality
  - Managing Scope Creep
  - Managing the Budget
  - Managing Risk
  - Managing Communications

- For each activity, there are MIS tools that can facilitate the management process. The two activities covered in this chapter were:
    - Controlling the Tasks
        - Project Plan
        - Detailed Execution Checklist
        - High Level Execution Status
        - High Level Test Case Status
    - Assessing Quality
        - High Level Code Delivery Tracking Report
        - Detailed Test Reports
        - Summary Test Reports
        - Ad Hoc Reports
- There are some common categories of Issues and solutions to those Issues for each major testing activity:
    - Major Issues Categories
        - Controlling the Tasks
            - People
            - Hardware/Software
            - Test Methodology/Status
            - Management
            - Other
        - Assessing the Quality
            - Serious Design Flaws
            - Large Number of Defects
            - More Critical/High than Medium Low Defects
            - High Number of Defects from One Coder
            - Large Number of Past Due Deliverables
    - Solution sets vary for each issue within each category, however the solutions typically are the result of a combination of adjustments across the following categories:
        - People
        - Scope
        - Hardware/Software
        - Test Method and Calendar
        - Management
    - Early Defect detection is critical to keeping the Test Calendar on schedule

XII. MANAGING TEST PROJECT ADMINISTRATION

In addition to managing the overall Test Processes, there are several categories of Project Administration that must also be managed. The previous chapter posed seven questions, and this chapter will address the remaining three questions:

- Has there been any Scope Creep?
- Is the cost within the allocated Budget?
- Are there pending Risks that could impact the timeliness, quality, cost and usefulness of the product?

Processes that can be used to answer these questions are provided in the following sections.

Scope Creep Management

There are two primary types of Scope Creep that concern the Test Manager.

- Functional Scope Creep
- Test Case/Test Procedure Scope Creep

Either of these two categories could disrupt the Test Calendar, and the Test Manager's subsequent ability to meet the planned deadlines. When both of these events occur at the same time and go unchecked, it is very difficult to recover.

Functional Scope Creep

Since it is the Test Manager's direct responsibility to oversee the writing of the Test Cases and Test Procedures, it is imperative that he/she is aware of any Functional Scope changes. To mitigate the Risks of delays, there should be a formalized process to communicate modifications to the Function and the Design.

All changes to Scope, large and small, should go through a review process that requires approvals from all of the affected parties. This typically involves:

- Evaluating the impact of the change on the Testing Scope
- Identifying the reasons the change is needed
- Conducting a cost/benefit analysis
- Assessing the result of the change on the overall Application

Once this information is assembled it should go through a management review and a formal approval process before the need for the change is accepted.

A Change Control Committee that consists of representatives from the Business Area, Development Team, Test Team, and other impacted areas most often conducts this review process. Usually, if the cost is over a specified amount, the request, in addition to being approved by the Change Control Committee, must also be approved by senior management, either in the form of a Steering Committee, or whatever other forum is dictated by the organization.

> **Authors' note:** A process to identify Design and Functional changes should be agreed upon at the beginning of the Project.

Interestingly enough, it is actually not always these larger items that tend to cause the most problems. Significant modifications are typically more obvious to a larger group of people, and most of them do go through the formal analysis and approval processes. As a result, the impact is usually well known and understood. It's when that review doesn't happen, on large or small items, that the Test Calendar is affected without any warning. Unfortunately, even when a process is in place, items can still be missed, and the cumulative effect of these missed items can affect the schedule.

If a process does not exist, the Test Manager should establish one in order to know when a change occurs, and ideally participate in the decision process to determine the impact on testing. Although it should not be part of the Test Manager's job to establish the process, if a process is not in place it could impact the testing schedule. To avoid this, the Test Manager should insist that the process be established.

Awareness, a good process, and good MIS help to mitigate the Risk of a delay and at the same time provide the Test Manager with enough information to manage through the difficulties when the inevitable Scope Creep does occur.

Test Case Scope Creep

In addition to Functional Scope Creep, it is possible for the number of Test Cases to begin to grow over time, without a conscious decision to do so. As the Code and circumstances evolve, new tests are required. Sometimes they should replace old tests and sometimes they are entirely new. And, occasionally a Developer or an End User will ask a Tester to perform an extra test.

> **Authors' note:** Even with a Change Management Process, some Scope Creep will inevitably occur. The Test Manager should be prepared to respond to the changes.

The Test Manager must be alert to the addition of Test Cases and ensure that no testing is being conducted that does not support the overall test objectives. This is covered later in this chapter.

Define Meaningful MIS

The Test Manager should define MIS that will provide information on the Functional Scope Creep, as well as the Test Case or Test Procedure Scope Creep. Usually a combination of some existing Reports coupled with a few additional ones should provide enough information for the Test Manager to successfully manage the Issues that are caused by both situations.

As with any potential problem, it is important to recognize early that there is a problem. The following section provides insight into how to identify potential problems, and some thoughts on how to make the necessary adjustments to correct them.

Functional Scope Creep

Awareness

Ideally, through the review process mentioned above, the Test Manager will be aware of any modifications to the Functions. Once approved, the new and/or modified Functions should be:

- Added to all of the relevant Control Documents
- Factored into the Test Calendar
- Used to prepare the necessary test material

If this process is followed, and the Test Manager agrees that there is enough time to get the testing accomplished, managing the newly added Functions should become a routine part of the overall management process.

However, since not all of the modifications are captured through the formalized process, the Test Manager must find a way to identify when an addition or modification occurs in a less obvious manner. This usually is accomplished in a variety of ways, and each manager must establish a process that works best for each test effort. Some of the general steps a Test Manager can take to Control this are:

- Ensure that each Test Team member understands the importance of informing the Test Manager when any change to Scope has been identified. Often, there are side agreements between a Developer and a Tester to squeeze in some additional testing. This might not seem like a problem to the individual Tester, but it could be a problem from the Test Manager's perspective.
- Even though the modifications may not formally be agreed on and documented, information about the new Function is usually contained in various Control Documents for review purposes. Very often it is in either a Development status meeting or a joint Project status meeting when a Tester or a Test Manager finds out that a new Function has been added. Even when it is identified late in the testing process, the Test Manager must to go through the assessment process and determine the impact on testing.

If the Test Manager does not think that the testing of the new Function can be accommodated with the existing resources and timeframes, this information must be communicated to the appropriate parties. Of course, the Test Manager should do everything possible to accommodate the Change if it is truly needed; and all of the possible adjustments that could enable the additional work to fit into the Test Calendar should be explored. If the Test Manager ultimately decides that the new Function cannot be accommodated, the Project Management Team and possibly the senior managers should be told.

Additionally, the Test Manager should periodically compare the Development Project Plans to the Issue Tracking Reports and Test Project Plans. This is just one more precautionary measure to ensure that there are no surprises.

Other Tools

Once a new and/or changed Function has been identified, agreed on, and incorporated into all appropriate Documentation, it will be important for the Test Manager to keep a running list of the modifications and their impact. Inevitably, the Test Calendar will get off track because of one or more of these modifications. It will be in the Test Manager's best interest to have Documentation supporting the reasons for the possible delays.

> **Authors' note:** The Test Manager should keep a detailed list of the modifications that are added after the testing starts.

As with much of the MIS, the format and content can vary as long as it documents the important facts; what Function was added, who approved what and when, did things go according to Plan, and if not, what was the impact?

This Document should be updated as the reality of the testing unfolds. In some instances, the Planned adjustments work as anticipated and the Test Calendar remains on track. In other instances, the adjustments do not have the expected outcome, and the Test Calendar is affected. This Document will provide the Test Manager with information about the impact of the modifications, and will serve as source information to make a determination as to whether or not the impact of any of them will seriously affect the overall Implementation date.

The following is an example of a Report that could provide the Test Manager with this information. This Report could be created from any one of the base Control Reports that provide a summarized View of the Functions.

	FUNCTIONAL ADJUSTMENT TRACKING SUMMARY								
	Included in Original Requirements	Approval Date	Approved By	Code Delivery Date Plan	Code Delivery Date Actual	Test End Date Plan	Test End Date Actual	Comments	
Screens									
Log on	Y								
Trade Entry	Y								
Trade Validation	Y								
Modify Trade	Y								
Cancel Trade	Y								
Reverse Trade	Y								
Reports-Ops									
Operational									
Daily Trade Activity	Y								
Cancelled Trade	Y								
Failed Trade	Y								
Late Trade	N								
Price Tolerance Report	N								
Reports-Client									
Daily	Y								
Monthly	Y								
Annual	Y								
Ad hoc	N								
Transaction Types									
Buy	Y								

Sell	Y								
Buy with accrued interest	Y								
Sell with accrued interest	Y								
Interest	Y								
Cash Receipt	Y								
Cash Disbursement	Y								
Free Receive	Y								
Free Deliver	Y								
Revenue	N								
Expense	N								

Figure 113 – Adjustment Tracking Summary Report

Test Case/Test Procedure Scope Creep

Awareness

If the testing effort has taken longer than planned, one of the things to consider is whether or not additional Test Scenarios have contributed to the delay. Since the Test Calendar is based on the time it takes to Plan and Execute Test Cases and Test Procedures for each Function, increases in the number of tests could be one part of the cause.

Test Scope Creep is not abnormal since it usually occurs as more is learned about:

- The Application itself
- How the business will be using the product
- Anything that was overlooked in the initial Planning

Because of this, the Test Manager should have allocated some time in the Test Calendar for all forms of Scope Creep; however, if Scope Creep occurs and the Test Calendar does not include sufficient time, the testing could move off schedule. In order to respond, the Test Manager must know when and why the Test Scope is changing, and must make the adjustments necessary to keep the testing on schedule.

Therefore, before Scope Creep occurs, the Test Manager should take steps to prevent it. Some of these steps are:

Test Case Repository

Ensure that all Tests Cases and Test Procedures are documented and numbered in a Test Library. If the Test Case information is maintained centrally, different Reports could be produced to highlight when new cases have been added.

Total Test Cases

One of the simpler Reports could be a summary that is produced by date. This would give the Test Manager an overview of how many Test Cases were added, and when. Using this Report, the Test Manager must ascertain what caused the additional tests, and determine whether adding them will seriously affect the Test Calendar. If the Test Calendar will be impacted, the Test Manager must identify the adjustments that could be made to bring it back on track.

Test Case Tracking Report

If this Report shows that more Test Cases are Executed than Planned, it might be an indication that Test Cases have been added outside of the Plan. It could also just mean that Test Cases were shifted from one day to the next, with no net increase in the number of cases.

Issues Without a Test Case Number Report

A Report could be produced that highlights Issues that were entered without a Test Case number. This could be an indication that tests are being performed that have not been documented, and probably were not Planned.

Other Tools

The Test Manager should also have a Report that includes all of the Test Procedures that have been established. This will help the Test Manager see emerging patterns that might require a different kind of remediation. One version of this Report should be sorted by Business Function, and another version should be sorted by Tester name.

Business Function View

The Business Function View highlights which Functional area had the most modifications, and prompts the Test Manager to examine the potential reasons, such as:

- Bad Requirements (Incomplete, inaccurate or poorly written)
- Addition of new Business Scenarios
- System not functioning exactly as anticipated

Tester View

Similarly, if the Sort by Tester indicates that there is a high percentage of new Test Procedures established by one particular Tester, the Test Manager should review this information with the Tester to understand what is causing the numbers to increase or decrease, which could indicate:

- Incomplete Test Scenarios or Test Cases
- The Application Functionality was not in the Functional Requirements Document

Once the answers to these questions are known, the Test Manager must decide whether any adjustments to the Plan are warranted, and if so, determine whether or not those adjustments will be sufficient to stay on schedule.

Reviewing the MIS

There is no industry Standard that defines how frequently this type of information should be reviewed. In the case of Scope Creep, daily reviews are probably not warranted; however, as with the review of all of the other MIS, the

Test Manager needs to determine the frequency. A possible review scenario might look as follows:

POSSIBLE SCOPE CREEP MIS REVIEW SCHEDULE			
REPORT TYPE	REPORT	WEEKLY	MONTHLY
Functionality	Compares Development Control Documents to test Control Documents		X
	Functionality Scope Control Log		X
Test Case / Test Procedure	Test Cases Created by Date Report	X	
	Test Case Execution Tracking Report	X	
	Issues Without Issue Number Report	X	
	Test Case/Test Procedure Control Log		X

Figure 114 – Sample Scope Creep Schedule

Regardless of the review schedule that is established, the nature of testing is that it can go from good to bad very quickly, and by reviewing these Reports with some regularity the Test Manager can reduce the chance of hidden problems that could cause serious damage.

Adjusting the Plans

Once it has been determined that there has been some Scope Creep that is causing a problem, there are a number of actions that can be performed to keep the Test Calendar on schedule. Before discussing these possible solutions, we will first review the steps that could be taken to prevent the situation from getting to the point where it is jeopardizing the Project.

To help reduce the likelihood of Scope Creep:

- Ensure that pre-Coding reviews are performed. These reviews dramatically reduce the number of missed Functions, misunderstood Functions, Designs that will not support the full Scope of Business Requirements, missed Test Cases, incomplete Test Cases, etc.
- Ensure that the business areas and the Test Team are represented during these reviews. The education of the Testers during this time is invaluable, as is the contribution from both the business areas and the testing staff.
- Ensure that there is a Functional Change Control procedure in place to reduce the chance of non-critical Functions being added. It also ensures that all of the parties understand the potential impact to the Project. Additionally, these review sessions may result in some very creative solutions.
- Ensure that MIS is in place to help identify potential problems as early as possible.
- Ensure that there is a structured testing methodology in place, such as the process described in this book. Having this structure in place facilitates the:
  - Organization of the test material
  - Ability to easily retrieve relevant information on the status of the Planning, Execution, and quality of the testing

Once Scope Creep has been identified as a problem, the Test Manager needs to evaluate every possible combination of adjustments to determine if any one or more of them will help keep the Test Calendar on track. The potential solutions were discussed in Chapter VI.

If these adjustments do not fully correct the situation, and if there is a large, complex new Function that cannot be satisfactorily tested before the Implementation deadline, the Test Manager will have to determine the answers to the following questions:

- Can the software be put into Production some time shortly after Implementation?
- Can the software be phased-in?
- Can any/all of the Medium/Low Defects be deferred?
- Are there any other Functions similar in size and complexity that can be deferred to allow time for the new Function to be tested?

If the answer to these series of questions is no, the test end date will probably have to be extended. And while this will not be an easy discussion, it needs to happen. The only thing worse than missing a target date, is moving an untested Function into Production that could become a liability for the business.

Financial Management

Once the testing effort has started, it is the Test Manager's responsibility to ensure that the Budget stays on target. Four areas that are within the Test Manager's Control that can significantly contribute to overspending are:

- People (Number of people and cost per person)
- Travel and Entertainment
- Hardware
- Software

In general, Budget overspending occurs because:

- Some activities were not Planned for at all
- Some were underestimated as a result of poor Planning
- Some were underestimated as a result of changing Project requirements that evolve as the Project progresses

It is important for the Test Manager to have enough information to track each of these actual expenditures against the Plan, and either make adjustments when necessary and possible; or when not possible, communicate the status to the appropriate parties.

- Breaks to the salary Plan typically constitute the largest portion of the variance. Salaries can very quickly get off track for a variety of reasons, and those reasons could vary from day to day.
- The variances to Plan for T&E, hardware, and software licensing fees usually happen because they were not Planned, or because there were some hidden costs such as additional Maintenance Fees on the software that were not understood. These variances are often a one-time occurrence and do not usually need to be monitored on a daily/weekly basis.

In any case, it is important to have good MIS in place to track of each of these Components. As with all of the other activities that need to be reviewed, the key here is to uncover anything that could possibly get the Project off track as early as possible.

Develop Meaningful MIS

Every organization will have their own process to track Project costs and report on the status. Regardless of the process that is used, it will be important for the Test Manager to have a reporting mechanism that allows for periodic reviews of how the test dollars are being spent. The Report formats and Sorts, if not defined by the organization, should be prepared in a way that works best for the Test Manager.

The following is a sample of a number of different Report Views that could help identify potential problem areas. While these samples could be useful as a template, the Test Manager should define the new Reports (or use existing Reports) that will help him/her isolate the potential problem areas as quickly as possible.

Total Test Budget – By Expenditure Type

This View provides summary level information about each of the main Controllable expenditure types (Salaries, T&E, Hardware and Software), and allows the Test Manager to quickly see all of the variances to the Plan by category type.

This Report also provides summary level headcount information. Since salaries are very frequently the source of a variance, having headcount totals on the same Report can quickly highlight whether the variance is due to having more people on the Project than Planned; or if the headcount is on target, this Report could lead to an investigation of rates, overtime, etc.

The following is a sample Report that could be prepared to coincide with the staffing Plan that was provided in Chapter VII.

\	TOTAL TESTING BUDGET BY EXPENDITURE TYPE ($000)												
Week Numbers		1	2	3	4	5	6	7	8	9	10	11	12	Total
Salaries	P[40]	23.5	22.8	22.8	22.8	22.8	22.8	22.8	22.8	22.8	18.8	18.8	14.8	258.8
	A													
Travel & Entertainment	P				.5				.5				.5	1.5
	A													
Hardware	P				12.0									12.0
	A													
Software	P				6.0									6.0
	A													
Licenses	P		10.0											10.0
	A													
Total	P	23.5	32.8	22.8	41.3	22.8	22.8	22.8	23.3	22.8	18.8	18.8	15.3	287.8
	A													
	HEADCOUNT													
Employees	P	4	4	4	4	4	4	4	4	4	4	4	4	48
	A													
Consultants	P	4	4	4	4	4	4	4	4	4	3	3	2	44
	A													
Total Headcount	P	8	8	8	8	8	8	8	8	8	7	7	6	92
	A													

Figure 115 – Sample Total Testing Budget Report

While this level of detail could be all the Test Manager needs to manage the Budget, different Sorts of the same information can help to isolate individual problem areas. The following five Reports provide various Views of this information at different levels.

Total Salaries/Headcount - By Test Phase

This View provides salary information Sorted by Test Phase for the entire test effort and allows the Test Manager to quickly see which Test Phases have the variances. A headcount summary is also provided to compare the headcount to dollars. This Sort also helps isolate the problem areas, and could be a time saver by allowing the Test Manager to focus in on the trouble spots.

---

[40] P = Plan, A = Actual

TOTAL SALARIES / HEADCOUNT BY TEST PHASE ($000)														
Week Numbers		1	2	3	4	5	6	7	8	9	10	11	12	Total
Test Manager	P	3.0	3.0	3.0	3.0	3.0	3.0	3.0	3.0	3.0	3.0	3.0	3.0	36.0
	A													
Administration	P	0	.8	.8	.8	.8	.8	.8	.8	.8	.8	.8	.8	8.8
	A													
Development														
Unit Testers	P	11.0	5.5	5.5	1.5	1.5	1.5	1.5	0	0	0	0	0	28.0
	A													
Functional														
Testers	P	9.5	13.5	13.5	9.5	5.5	1.5	1.5	0	0	0	0	0	54.5
	A													
Business Integration														
Testers	P	0	0	0	8.0	12.0	16.0	16.0	13.5	9.5	1.5	1.5	0	78.0
	A													
User Acceptance														
Testers	P	0	0	0	0	0	0	0	5.5	9.5	13.5	13.5	11.0	53.0
	A													
Total Salaries	P	23.5	22.8	22.8	22.8	22.8	22.8	22.8	22.8	22.8	18.8	18.8	14.8	258.3
	A													
HEADCOUNT														
Test Manager	P	1	1	1	1	1	1	1	1	1	1	1	1	12
	A													
Administration	P	0	1	1	1	1	1	1	1	1	1	1	1	11
	A													
Development														
Unit Testers	P	4	2	2	1	1	1	1	0	0	0	0	0	12
	A													
Functional														
Testers	P	3	4	4	3	2	1	1	0	0	0	0	0	18
	A													
Business Integration														
Testers	P	0	0	0	2	3	4	4	4	3	1	1	0	22
	A													
User Acceptance														
Testers	P	0	0	0	0	0	0	0	2	3	4	4	4	17
	A													
Total Headcount	P	8	8	8	8	8	8	8	8	8	7	7	6	92
	A													

Figure 116 – Sample Total Salary Report

Travel and Entertainment

Planning for this expense is often overlooked and/or underestimated. Some of the key considerations when establishing the Budget and tracking these costs are:

Travel/Lodging costs for traveling for:
- Testing
- Training
- Conversions

Accommodations for:
- Cabs for late night travel
- Conversions
- Hotels

Food for:
- Accommodations
- Celebratory dinners/parties
- Overtime meals

> **Authors' note:** In a large Project when there is vendor outsourcing or support involved, T&E is often either not clearly defined in the contract, or the implications are not fully understood. The Test Manager must do a thorough job of estimating these expenses at the beginning of the Project. Once the details have been identified and Budgeted, a simple Report that compares the Planned to actual expenses will prevent this from getting out of Control. The key is to be aware of any variance.

Hardware/Software

More often than not, the Issue with overspending in this category is a result of not Budgeting at all for the needed equipment or software licenses. An evaluation should be performed on the following items at the start of the Project:

- Hardware
  - PCs
  - Printers
  - Servers
  - Data Center charges (if charged back to the Test Team)
- Software
  - Project Management
  - Issues Repository
- Test Automation Tools
  - Other general business Applications (i.e. Access, Excel)
- Licenses
  - Copies of the software required
  - Site license or individual licenses

Similar to the salary and headcount Reports, once the Budget has been prepared, a Report that compares the actual to the Planned spending on these individual Components will help monitor the spending. This Report could be prepared and tracked at the Project level, or if warranted, at the Test Phase level. Often, these charges are not Test Phase specific, and monitoring at the Project level should satisfy the MIS need.

In addition to the information provided at the Project level, it often helps to have more detailed information for each Test Phase. The following two Reports reflect various Sorts of the same core information with details by Test Phase. The following Report provides information by resource type.

Salaries/Headcount - By Resource Type – Within Test Phase

SALARIES / HEADCOUNT (FUNCTIONAL PHASE) BY RESOURCE TYPE ($000)													
Week Numbers		1	2	3	4	5	6	7	8	9	10	11	12	Total
Employees														
Full Time	P	1.5	1.5	1.5	1.5	1.5	1.5	1.5	0	0	0	0	0	10.5
	A													
Part Time	P	0	0	0	0	0	0	0	0	0	0	0	0	0
	A													
Total Employees	P	1.5	1.5	1.5	1.5	1.5	1.5	1.5	0	0	0	0	0	10.5
	A													
Consultants	P	8.0	12.0	12.0	8.0	4.0	0	0	0	0	0	0	0	44.0
	A													
Total Salaries	P	9.5	13.5	13.5	9.5	5.5	1.5	1.5	0	0	0	0	0	54.5
	A													
HEADCOUNT													
Employees														
Full Time	P	1	1	1	1	1	1	1	0	0	0	0	0	7
	A													
Part Time	P	0	0	0	0	0	0	0	0	0	0	0	0	0
	A													
Total Employees	P	1	1	1	1	1	1	1	0	0	0	0	0	7
	A													
Consultants	P	2	3	3	2	1	0	0	0	0	0	0	0	11
	A													
Total Headcount	P	3	4	4	3	2	1	1	0	0	0	0	0	18
	A													

Figure 117 – Sample Salaries By Resource Type Report

Once it has been established that there is a variance to the Plan, this Report helps to further identify the cause. This View can identify whether the problem was with the consultant Budget or perhaps is a more general problem with overtime.

Some possible reasons for a variance are:
- People stay longer than Planned
- People come in at a higher rate than Planned
- People scheduled with earlier end dates are needed to stay longer, etc.

Regardless of the format, it is critical to keep track of the people and the money that is paid to employ them. A Report like this keeps the numbers constantly available for use by the Test Manager.

Salaries/Headcount - By Function – Within Test Phase

Another helpful Report contains salaries and headcount that is Sorted by Function. This View allows the Test Manager to see if a Functional area is overspending. This Report can most easily be produced if a time tracking Application is used which captures the actual time that is spent on specific activity types.

		\multicolumn{12}{c}{SALARIES / HEADCOUNT (FUNCTIONAL PHASE)}
		\multicolumn{12}{c}{BY FUNCTION ($000)}
Week Numbers		1	2	3	4	5	6	7	8	9	10	11	12	Total
GUI	P	4.0	4.0	4.0	0	0	0	0	0	0	0	0	0	12.0
	A													
Reports	P	4.0	4.0	4.0	4.0	0	0	0	0	0	0	0	0	16.0
	A													
Interfaces	P	1.5	5.5	5.5	5.5	5.5	1.5	1.5	0	0	0	0	0	26.5
	A													
Total Salaries	P	9.5	13.5	13.5	9.5	5.5	1.5	1.5	0	0	0	0	0	54.5
	A													
\multicolumn{15}{c}{HEADCOUNT}
GUI	P	1	1	1	0	0	0	0	0	0	0	0	0	3
	A													
Reports	P	1	1	1	1	0	0	0	0	0	0	0	0	4
	A													
Interfaces	P	1	2	2	2	2	1	1	0	0	0	0	0	11
	A													
Total Headcount	P	3	4	4	3	2	1	1	0	0	0	0	0	18
	A													

Figure 118 – Sample Salary by Function Report

Any one, or a combination, of these Reports can be useful to the Test Manager to manage the Budget. To prepare the Reports, it is necessary to track the details at a detailed level either on an automated time tracking Application, manual spreadsheets, or a combination of the two. If the details are regularly tracked, the Test Manager can accumulate them for review. While the format, content and Sort is not really important, it is important to have some way to find the source of the variances so the Test Manager can make the necessary adjustments.

Reviewing the MIS

In general, for a Large Project, the status of the Budget should be reviewed on a monthly basis. For illustration purposes, the Project depicted in Chapter VII was a short-term 12-week effort, and subsequently, again for illustration purposes, the staffing and related MIS was prepared accordingly.

More realistically for a Large Project, the Budget, even if Planned on a weekly basis would be rolled up and reviewed on a monthly cycle. Regardless of the frequency of the review, the Test Manager should look at both the totals for that period as well as the YTD totals. It is very often the case that the numbers are over the Budget for one month, and then under the next, frequently netting out to be on Budget. The review of the status of the current period and YTD categories will give the Test Manager an indication of how things are compared to the overall Plan.

Also, depending on the frequency of the senior management review schedules, the Test Manager might have to prepare a separate set of Reports that reflect a collapsed, higher level View of the Budget status. As long as the underlying data is tracked, there should be no problem satisfying most reporting requests.

Reviewing the Plans

As with any deviation from Plan, the important thing is to understand what is causing the variance, fix it, and make the necessary adjustments to bring the Plan back in alignment. The Reports previously described will highlight those variances, and provide some insight on the cause. The solutions that are used to rectify the situation will vary depending on the nature of the problem, and the status of the other activities that are underway.

If, for example, more money was being spent on testing the Interfaces than Planned, but less on testing the Reports, it might be that the net effect will keep the Budget in line. However, if there is a net overspend, an assessment needs to be made to determine what adjustments could be made to bring the actual spending back in line with the Plan. The chart that follows identifies some of the more common reasons for financial variances to occur, and possible steps to take to get the budget back on track.

Issue Category	Issue	Possible Solution (s)
People	Need to bring in more people than Planned	Use business people.
		Supplement with lower priced temps/school interns to do the less critical work.
		Attempt to get testing done sooner to be able to let resources go earlier than Planned to make up the dollar difference.
	Need more skilled /higher price people than Planned.	If the Test Calendar allows, Train less experienced people.
		Use more experienced business people that have an interest and proclivity towards Project work.
		Realign Tasks to give more complex work to existing senior people. If necessary, bring in lower priced people to supplement the experienced staff.
		Attempt to get testing done sooner to be able to let resources go earlier than Planned to make up the dollar difference.
	Need to extend people longer than Planned.	Negotiate better rates with consultants. Often, consultants are willing to make this compromise for a longer assignment, and a better chance at possible future work.
		Supplement with employees/ business people instead of extending consultants.
	Paying consultants hourly vs. per diem as expected.	Negotiate different contract.
		Negotiate lower rate.
	Spending more on overtime than Planned.	Work in shifts to eliminate overtime.
		Supplement with business people.
Hardware / Software / Licenses	More equipment needed than Planned.	Borrow surplus from other areas in the organization.
		Share equipment, where possible.
		Work in shifts.
	Cost of equipment higher than Planned.	Shop for better deals (possibly outside of normal purchasing channels)
		Reevaluate whether fewer pieces of equipment could be used.
		Share equipment, where possible.
	Unexpected licensing fees	Limit use of software, where workable, to a smaller group of people.
Travel and Entertainment	More and/or more long termed travel than anticipated	Use apartments vs. hotels for longer term stays.
		Shop for better rates.
		Condense activities where possible into fewer, longer-term stays to save on travel costs.

Figure 119 – Spending Issues and Possible Solutions

These are some of the more common problems and solutions, and the Test Manager must evaluate each situation individually and determine the best combination of adjustments to make.

Risk Management

Ideally, the Project Manager should perform a full Risk Assessment for the entire Project, and an assessment of the testing activities should be one aspect of the review. The purpose of any Risk assessment is to:

- Identify those things that could either delay the Project, cause the Project to go over Budget, or potentially result in an Application that either failed to perform according to the Requirements, was unstable, or of poor quality

- Evaluate the likelihood of these things happening, and the impact to the Project if they did
- Prepare a Risk Mitigation Plan to avoid the highest rated items
- Manage and/or adjust the Plan when/if the Risk situation becomes a reality

The Test Manager needs to be concerned with all of these things on a smaller scale as they relate to the testing activities. If the organization does not mandate a format, categories and/or formulas for quantifying Risk, the Test Manager and the rest of the Project Management Team need to establish them. This should be an ongoing process where the Risks are identified, evaluated, and re-evaluated as the Project progresses. There are a number of approaches to managing Risk. One approach takes a look at it from three different perspectives:

- High Level Assessment of the organization's appetite for Risk
- Risk Analysis of all of the possible Risk factors
- Revalidation/Revision of the remediation Plans

Develop Meaningful MIS

High Level Risk Assessment

Very often, Risks are assessed first at a very high level first to determine the organization's level of concern for each of the Risk categories. A simple analysis might look as follows:

RISK CATEGORY	POSSIBLE EVALUATION
Missed Implementation Date	Can not recover if the Delivery Date is missed
Over Budget	Some flexibility; however, a significant variance will negate the cost/benefit analysis
Does not fully meet the Business Requirements	Application delivered must meet exact Business Requirement; however, there is a possibility of a phased Implementation if absolutely necessary
All of the required Functions are not delivered	Application can move into Production if there are tested workarounds in place for High, Medium and Low. Critical Issues must be resolved

Figure 120 – Sample High Level Risk Assessment

It might be that there is zero tolerance for a missed deadline because the Project is supporting, for example, a new Regulatory Requirement. This information, to the degree that the Test Manager knows it, is helpful in both the Planning stage, as well as throughout the Control process. In this case, if meeting the Project deadline was in jeopardy, the Test Manager should know there could be some flexibility in possibly bringing in additional people, and/or segmenting the deliverables. This information is not something that should necessarily be shared with every member of the Test Team, but it is something that would be extremely useful to the more senior members of the Project Management Team.

> "One of the tests of leadership is to recognize a problem before it becomes an emergency."
> Arnold Glasgow

Risk Analysis

Beyond the High-Level Risk Assessment, a more detailed Risk Analysis should be prepared for every possible Risk factor. This analysis can be prepared by using the following process:

- Identify Significant Risk Factors
- Assess/Quantify the probability of those Risks occurring
- Assess/Quantify the impact to the Project

Once completed, this information will help the Test Manager see the Risk factors that fall into the highest probability and highest impact category. These are the Risks that could potentially devastate a Project, and therefore must be managed very Closely.

There are a number of ways to prepare this analysis. One simple method is to give a value of High, Medium and Low to both the probability and impact factors. For illustration purposes, the following Matrix has been prepared using the Issues from the Test Issue/Solution Matrix (Figure 112). In all likelihood, many of these items would be part of the Risk analysis along with other potential Risk factors.

		MASTER RISK ASSESSMENT LIST						
		RISK FACTORS	PROBABILITY			IMPACT		
			H	M	L	H	M	L
I. People	a.	Testing staff not in place when testing starts		X		X		
	b.	Insufficient number of Testers		X			X	
	c.	Inexperienced testing staff	X				X	
	d.	Unfamiliarity of testing staff with the business			X		X	
	e.	Unfamiliarity of testing staff with the testing tools			X			X
	f.	Staffing turnover rate			X	X		
	g.	No compensation structure for overtime and weekends	X					X
	h.	Poor relationships between Testers, Developers and user			X			X
	i.	Uncooperative team players			X	X		
II. Hardware / Software	a.	Late Code						
	b.	Test Environment not ready on time						
	c.	Scope creep						
	d.	Poor quality Code						
	e.	Slow turnaround on Defect fixes						
		System Issues:						
	f.	- Response time						
	g.	- Availability						
	h.	- Significant Design flaws						
III. Test Status	a.	Test material not ready on time						
	b.	Poor test preparation and Execution						

Sample Responses

	c.	Inaccurate test estimates					
IV. Management	a.	Inadequate senior management support					
	b.	Disinterested managers of involved / impacted groups					
	c.	Poor Test Management					
V. Other	a.	Budget					
	b.	Audit					
	c.	Politics					
	d.	Size of the Project is larger than normal					
	e.	Complexity of the Project is greater than normal					
	f.	Development staff is unfamiliar with the technology					
	g.	Miscellaneous Business / Project Issues					

Figure 121 – Sample Risk Assessment Report

Every item will need to be evaluated for the probability of it occurring, and the cost to the Project if it does. Once the evaluations are complete, they can be mapped into the following Risk Matrix (Figure 115). The Risks that fall into the high probability/high impact category obviously represent the biggest Risk to the Project. Those falling into the low/medium probability, but high impact category are also troublesome. Even though less likely, if they do happen they represent as much Risk as the high/high.

The intent of this type of Matrix is to provide insight into those factors that need to be managed very closely. The overall objective for the Test Manager is to pro-actively manage the High probability/High impact Risk items in order to move them into the Low and Medium probability categories.

Each item should be entered into the appropriate cell using the numbering scheme that was defined on the Risk Assessment Report. The Test Manager should be concerned if he/she finds that there are many items in the High/High or even Medium/High categories. Minimally, the High probability/High impact items should have a strategy prepared to avoid them, and to deal with their impact if they occur.

RISK ANALYSIS MATRIX			
Probability	**Impact**		
	Low	**Medium**	**High**
High	1.g	1.c	
Medium		1.b	1.a
Low	1.e, 1.h	1.d	1.f, 1,i

Figure 122 – Sample Risk Analysis Matrix

Ideally, anything that has a High, and to a lesser degree, Medium impact should be seriously reviewed for ways to minimize the likelihood that they will happen, with consideration for the possible remediation steps that could minimize their impact.

Revalidation/Revision of the Remediation Plans

Once a Risk item becomes a reality (e.g., the Code is delivered late to the Test Team), the impact and possible solutions must be re-examined to see if any factors have changed from when the analysis was initially prepared. The Test Manager, along with the rest of the Project Management Team, will need to revalidate where the item falls in the Risk Assessment Report, and make adjustments accordingly.

Also, when a new Issue arises that has significant Risk potential that was not part of the original evaluation, it should be added to the Matrix for future evaluation, and remediation Plans prepared. At this point, the revalidated, revised and/or new remediation steps should get integrated into the rest of the Plans and managed by the Test Manager along with the other Tasks.

Reviewing the MIS

The Test Manager must be aware of potential Risk items on a daily basis by using the Tracking Report's review and prioritization process to identify serious Issues as they arise. Through a combination of the preparatory analysis work that was described above, and the daily/weekly Issue Tracking Review Meetings after the testing starts, a more complete Risk picture will emerge. Although there really is no industry Standard for a review schedule, a Risk Review Calendar could look as follows:

POSSIBLE RISK REVIEW CALENDAR			
	BEGINNING OF PROJECT	DAILY / WEEKLY	MONTHLY
High Level Assessment Review	X		
Issues Review		X	
Full Risk Matrix Review			X

Figure 123 – Possible Risk Review Calendar

Assessing and managing the Project's Risk factors is one of the more difficult roles of the Test Manager, or any manager. Experience plays a big part in seeing the critical path items and understanding their potential impact on the Project, knowing to constantly look for the Risks, and in being able to develop a Plan of action to resolve them. The tools reviewed here will be helpful, but they must be used along with a good level of testing experience, as well as an understanding of the product and business and common sense.

Adjusting the Plans

In reality, all of the adjustments covered under the Scope, Budget and Risk sections of this chapter, as well as the Tasks and Quality sections in the previous chapter, represent the possible solutions to the most likely Issues that could potentially be Risks to a Project. The combination of the suggested Control processes and Reports that are reviewed throughout this book should position the Test Manager to be better able to identify, evaluate and react to these Issues in a timely and meaningful way.

Communication Management

The last piece of the Controlling process is Communications. Depending on the size and nature of the Project, the audience, type and frequency of the Communications will vary. If a Communications Plan was established during the earliest stages of the Project, the stakeholders would likely have been identified, along with the type and frequency of the Communications. If a Plan was not established, the Test Manager will need to define the information requirements.

Stakeholders

Typically, in a Medium to Large Project, several groups of internal and external parties will be interested in the progress of the testing effort, including:

- Internal
  - Test Team
  - Project Management Team
  - Senior Management
  - Internal Auditors
- External
  - Clients
  - External Auditors
  - Media (Industry, Other)

Generally, information sharing between any one of these parties occurs either through a series of Meetings, or through formal and/or informal Documentation. Most often, the members of the Test Team are primarily involved in the internal Meetings, and the Project Manager either provides updates to and/or attends Meetings in support of the external stakeholders.

The following section reviews the general types of Meetings for the Test Team. Following the Meeting section is a review of the basic types of Documents that are frequently used in the process.

Meetings

The primary purpose of the testing related Meetings is to provide a forum for the exchange of information. It is during these Meetings that people give updates on the status of their own work, listen to the updates of others, attempt to work through roadblocks, and/or make decisions about next steps, etc. In Medium to Large Projects there is usually a hierarchical structure that includes the following types of routine Meetings:

- Project Team Meetings
- Project Management Meetings
- Senior Management Meetings

In addition to these routine Meetings, there are some special categories of Meetings that are discussed in other chapters, including:

- Sign-Off Meetings
- Project Closing Meetings
- Performance Reviews

These various types of Meetings are discussed in the following sections.

Project Team Meetings

The Test Team members usually will meet among themselves on a fairly frequent basis to review the daily and/or weekly testing progress. Included in this category are the individual:

- Test Team Status Meetings
- Issues Management Meetings

As with many of the Components of Project Management, there are no firm rules about who should meet, how frequently they should meet, and what should be covered in those Meetings. As such, the following Meeting schedule represents one out of the many approaches that could satisfy a Project's information needs.

Test Team Status Meetings

These Meetings are typically run by and attended primarily by the Test Team, and are held with varying frequencies. One approach to their frequency is the following:

Daily

Most members of the Test Team will likely be involved in daily Meetings where the day's detailed Tasks are reviewed and the next day's Tasks are discussed. Depending on the size of the Project, the Test Manager may manage this Meeting or in the case of a Large Project the Section Managers may hold their own Meetings. Members of the Development Team and the End Users might attend some or all of these Meetings.

The Execution Checklist is often the Document that is used to guide these daily Meetings. Typically, the progress of all of the tests and other activities that are planned for the day are reviewed, including the Execution status of the Test Cases, as well as the timing of the on-line and/or batch processes.

This daily Meeting is usually heavily focused on making the decisions that need to be made to get through a single Test Day, i.e., what Files to run and when, what processes to run, and what to do about any events that are affecting the tests.

Weekly

In a Large Project, the Test Manager will normally meet with his/her direct Reports alone, probably on a weekly basis. The purpose of this Meeting is to get a higher-level update from each of the Section Managers regarding the progress of their individual areas.

It is typically during this type of Meeting where concerns are discussed about impediments to the overall deadlines/and or about the quality of the product, etc. These sessions are usually held to identify solutions, and establish a joint Plan of action to get things back on schedule. These steps might involve moving people temporarily from one section to another to get through a particularly difficult time, or if need be, making changes to the Test Calendar.

Ad Hoc

There are a variety of ad hoc meetings that are held throughout the life of any Project.

- There is typically a hand-off that occurs when a new software release is delivered, which should include the Code and the release notes that document the contents of the Code. A Meeting is typically scheduled to review the contents of the release to ensure that all interested parties fully understand what is delivered. This Meeting is attended by the Technology and Test Managers, but may include other interested parties.

- Also, as the need arises, any number of ad hoc Meetings can be arranged to review Critical/High Severity Issues or serious Risks that cannot wait for a regularly scheduled Meeting.

Issues Management Meetings

The review and management of the Issues Tracking Report usually involves several steps.

Establish the Severity Levels

The participants in this Meeting must review the Issues and either evaluate the Severity levels that were assigned by the Test Team, or assign the Severity levels for the first time. Depending on the stage of the Project, this process would probably take place on a weekly basis, or, as the deadline approaches, on a daily basis.

Review/Prioritize the Issues

This Meeting usually has representatives from the Test Team, Development Team, End Users and any outside vendors. The purpose of this Meeting is to:

- Review the status of existing Issues
- Review any new Issues
- Discuss and agree on the priorities of those Issues
- Evaluate the possible solutions to resolve the Issues
- Re-prioritize existing items to fit the more important new items into the schedule

The frequency of this Meeting varies by need, and the Test Manager usually determines that need.

Project Management Meetings

After the weekly update Meetings with the various Test Team Section Managers, a representative from the Test Team (usually the Test Manager) will meet with the larger Project Management Team. This weekly Meeting is usually led by the Project Manager and includes the Test Team Manager, the Development Manager, the IT Manager, End User representatives, Audit, Training, etc., and is typically held to exchange status information and to make decisions regarding the Project.

This type of Meeting is extremely beneficial since it is one of the few times where there is representation from every group involved in the Project, and where everyone is getting the same update from one source. It is at this Meeting where the Test Manager reports on the testing status that was previously provided by each Section Manager; and it is from these Meetings that the Project Manager gets a full picture of all of the key Project activities, along with the Issues that might have surfaced since the last Project Management Meeting.

Typically the updates from each area are sent in advance to the Project Manager in a pre-defined format, and a consolidated Report is distributed and reviewed during the Meeting. Although the format and content of this material will vary from organization to organization, whatever format the information is in should provide the Project Manager with a concise View of the:

- Status of the key milestones
- Current problems and recommended solutions
- Potential Risks and alternatives

Senior Management Meeting

Depending on the organization, the Test Manager may or may not accompany the Project Manager to the Senior Management Meetings. The focus and frequency of these Senior Management Meetings vary; however, there will always be a need to keep the Senior Managers informed of the Project's status. It most cases, it is the Project Manager, not the Test Manager, who will provide the update, although in some organizations, the senior managers of each of the disciplines (i.e., Development, Testing, Training, etc.) provide their own updates.

In many cases this is either a weekly or a monthly update where an update is provided on all of the relevant performance indicators for Task completion, quality, cost and Risk. In some organizations, the cost and Risk factors are reviewed less frequently.

Documentation

There are several different types of Documents that are prepared throughout the life cycle of the Project. From a testing perspective, there are three main categories of Reports that are produced to either inform people of the objectives of the testing effort, or to report on the status of the testing activities as the Project progresses. These categories are:

- Planning Documents
- Meeting Documents
- Status Reports

Each of these categories is described in the following sections.

Planning Documents

The Master Test Plan, as described earlier, consists of the Summary Test Plan, Detailed Test Plans and the Test Project Plan. These Documents serve many purposes.

- The Summary Test Plan[41] is typically intended to inform a broad audience of the:
  - Primary objectives of the testing effort
  - Testing approach
  - Estimated resources required
  - Projected timeframes for each Test Phase
- The Detailed Test Plans[42] are intended for a smaller audience who are interested only in certain aspects of the testing effort, and typically contain the working Documents that the Test Team will use to Execute and monitor their testing activities. These Plans are written for each Test Phase and contain the:
  - Details of the tests to be Executed
  - Detailed Plans for how each Test Day will be conducted

  The people most interested in the Detailed Test Plans are the Test Team members themselves, along with the Development Team who typically work very closely with the Test Team as they Execute their Plans.
- The Test Project Plan[43] is a tool that is primarily used by the Test Manager to manage the testing effort and is often used as a source of information for Status Reports and /or Meeting Documents.

Meeting Documents

Project Team Meetings

The format and content of the Meeting Documents vary from Meeting type to Meeting type; however, some of the basic Reports frequently used are:

- Daily Test Team Meeting
  - Execution Checklist
  - Selected Sorts/views of the Issues Tracking Report
- Weekly Test Team Meeting
  - Selected Sorts/views of the Project Plan
  - Test Calendar
  - Selected Sorts/views of the Issues Tracking Report
- Weekly Project Management Meeting
  - Integrated High Level Update (possibly including a milestone View of the Project Plan)
  - Selected Sorts/views of the Issues Tracking Report

Senior Management Meetings

The Documents for these Meetings are usually manually prepared and represent a consolidated View of the status of each of the Project's

---

[41] Chapter VI
[42] Chapter VII
[43] The Test Project Plan is reviewed in Chapter VIII

performance indicators. Typically, the Test Manager submits input to the Project Manager that will be integrated with each of the other Project Components. Performance indicators are typically reported for the following areas:

- Development
- Testing
- Training
- Conversions
- Quality
- Risk
- Cost

Testing Status Reports

Testing Status Reports are typically prepared by the Test Manager and distributed to various interested parties. Depending on the audience, the format and content may vary; however, there are generally two different types of Reports that are prepared:

Plan to Actual

This type of Report compares where each testing activity stands against the Plan.

	FUNCTIONAL TEST STATUS REPORT AS OF 09/01/02		
	SCHEDULED COMPLETION DATE	ACTUAL COMPLETION DATE	PERCENT COMPLETE
Test Preparation	07/29	07/20	100%
Test Execution – Day 1	09/11		75%

Figure 124 – Sample Test Status Report (Plan to Actual)

Accomplishments

This type of Report provides a summary of the testing accomplishments for a specified period of time.

TEST PHASE	ACCOMPLISHMENT
Functional Test	Completed Predicted Results for Days 1-5 - *10/31-12/31*
	Executed 150/200 Trade Entry Test Cases for Day 1 - *10/31*
	Completed validation of 50/75 Client Statements for Day 1 – *10/31*
	Re-tested 40/40 critical Issues for Day 1 – *10/31*

Figure 125 – Sample Test Accomplishment Report

Test Status Reports must be produced on a regular basis. They inform a wide audience about the progress of the Project, and help areas outside of the immediate Test Team understand the activity that is going on around them. They also serve as a validation of the work that is accomplished, and are a good historical source of information when managers are inevitably asked to produce any number of ad hoc Reports.

Document Repository

The Project Team should use a centralized repository to store all of the Project Documentation. When maintained properly, a Document Repository allows the latest versions of the relevant Documents to be accessed easily.

> **Authors' note:** A central Document Repository is essential for effective Project management.

One of the biggest problems in large Projects with many Documents, and multiple versions of those Documents, is that people can work with old and/or different versions. Storing the information electronically and centrally helps to mitigate that Risk significantly. It also makes it easier to produce regularly scheduled or ad hoc Test Status Reports.

Although it is time consuming, and somewhat costly to maintain, in the end it is more time consuming and costly for this information to be stored in multiple locations, and to potentially run the Risk of people working off of different versions.

Some of the key Documents that should be stored centrally are:

- Overall Project Plan
- Requirement Documents
    - Business Requirements
    - Functional Requirements
    - Technical Design Requirements
    - Data Design Specification

> **Authors' note:** Pay now to set up a central Document Repository, or pay more later when you've tested against the wrong Documents.

- Test Documents
    - Master Test Plan
        - Summary Test Plan
        - Detailed Test Plans
        - Test Project Plan
    - Business Scenarios
    - Test Cases
    - Test Procedures
        - Expected Results
        - Actual Results
- Conversion Documents
- Project MIS
- Meeting Minutes

The Document Repository does not need to be fancy or complex. Many people use either a Lotus Notes Database or a repository within their automated test tool if it has one. The key is to have the information stored in an electronic library where it can be easily accessed (in read-only mode) by many users, and to have a process in place to maintain it.

Additionally, there are other Documents related to the Project that should also be stored and accessible only by the Test Manager and possibly a few select members of the Project Management Team. These Documents include items such as:

- Budget Reports
- Personnel Related Reports
- Risk Analyses

Summary

The key points in this chapter are:

- In addition to conducting the tests, the Test Manager must also manage four other areas:
  - Managing Scope Creep
  - Managing the Budget
  - Managing Risk
  - Managing Communications
- MIS should be prepared to help manage the two primary types of Scope Creep, including:
  - Functionality Scope Creep
    - Comparison of Development Control Documents to Test Control Documents
    - Functionality Scope Creep Log
  - Test Case Scope Creep
    - Test Case Scope Creep
    - Test Case Repository
    - Test Case Tracking Report
    - Issues Tracking Repository
- There are four key Budget categories that are within the Test Manager's control:
  - People
  - Travel and Entertainment
  - Hardware
  - Software
- There are several type of Reports that can be used to manage the Budget, including:
  - Total Test Budget – By Expenditure Type
  - Total Salaries/Headcount – By Test Phase
  - Salaries/Headcount – By Resource Type – Within Test Phase

- Salaries/Headcount – By Function – Within Test Phase
- There are four steps in the Risk Management process:
  - Identify what could potentially impact the timeliness/quality of the deliverable
  - Evaluate and quantify the likelihood of these things happening
  - Prepare a Risk Management Plan
  - Manage the Plan
- There are two primary groups of stakeholders that the Test Manager needs to communicate with:
  - Internal – Project Management Team, Test Team, Senior Management, Internal Auditors
  - External – Clients, External Auditors, Media
- The content, frequency and participants in the Project Meetings vary. One possible combination is:
  - Project Team Meetings
    - Test Team Meetings
      - Daily
      - Weekly
      - Ad Hoc
    - Project Team Meetings
      - Weekly
    - Senior Management Meetings
      - Weekly
      - Monthly
- There are three key types of Documents:
  - Planning Documents
  - Meeting Documents
  - Status Reports
- It's important to have a centralized Document Repository to store:
  - Project Plans
  - Requirements Documents
  - Test Documents
  - Conversion Documents
  - Project MIS
  - Meeting Minutes
  - Budget Reports
  - Personnel Related Reports
  - Risk Analyses

XIII. MANAGING TESTING AUTOMATION

Regardless of whether testing is conducted manually or with automated tools, there are two overall testing approaches that are based upon the software Design concept of "input-process-output."[44]

White Box Testing

In the White Box Testing approach, Developers examine the processing logic that occurs inside the Application. White Box Testing focuses on evaluating the Code to trace the internal logic of the Program, and the Developers are concerned with the Execution of all possible logical flow paths through the Program. In their Unit Testing, the Developers are more concerned with 'how it is done' than 'what is done.'

Examples of White Box Testing are:

- Testing branches and decisions in Code
- Tracking a logic path in a Program
- Tracking data retrieval and posting to files

> **Authors' note:** White Box Testing is "how it is done."

Black Box Testing

In the Black Box Testing methodology, the Testers look at the Application the same way as the End Users will see it once the Application is in Production, and are concerned with inputs and outputs. Black Box testing uses the activities that are described in the Functional Requirements Document and is used in each Test Phase after the Developers deliver the Code. They are interested in 'what is done' by the Application, not with 'how it is done.'

The Testers create tests that replicate the Conditions that could be encountered when the Application's Functions are performed. All of the possible relevant input combinations are entered and the outputs are examined and compared to the Expected Results. Both valid and invalid inputs are used to test the Application.

> **Authors' note:** Black Box Testing is based on "what is done."

Some examples of Black Box Testing are:

- Enter an Automated Teller Machine withdrawal transaction and observe that the expected cash is dispensed
- Exercise an equity purchase Function by entering data and observe the results
- Cancel a funds transfer and observe that the transaction was cancelled
- Update a Client's address and observe that the new address is printed on the Client Statement

In this chapter, we will examine several of the techniques that are available to improve the testing process, which are ultimately a form of either Black Box or White Box Testing. These tests can be conducted manually or can be based upon any of the many available automated testing tools.

This chapter covers three topics:

---

[44] Input data, process the data, and see the output.

- First, the chapter presents the primary differences between manual and automated testing and presents a framework for making the cost/benefit analysis that can support that decision.
- The second topic identifies the tools that are available for testing Enterprise Applications and for Web-based Applications.
- Finally, the chapter includes a review of the different ways that the available testing tools can be used in each stage of the SDLC.

Over the past several years, the Development tools that help Developers quickly create Applications have dramatically improved Developer productivity. This has increased the pressure on Testers, who are often perceived as bottlenecks to the delivery of software products. Testers are being asked to test more and more Code in less and less time. To meet this demand, Testers need to dramatically improve their own productivity. Test automation is one way to do this. [45]

Authors' note: Test automation can dramatically improve Tester productivity.

Usually there is no clear line that identifies when to use either manual testing or automated testing. The very Smallest Projects can be manual and the Largest should be automated, but most Projects are in between, and have a mix of manual activity and automation. A technique for measuring the size of a Project was discussed in Chapter V.

Manual Testing

Manual testing is useful for Small Projects or in a very fluid situation. Managers can usually get more flexibility from a team of people than from a set of automated tools that must be carefully applied.

Staffing Requirements [46]

When staffing for a manual testing Project, the Test Manager will have to assemble a team of people with diverse skills in order to optimize the cost. The team will have to include:

- Test Section Manager
- Test Quality Assurance Analysts
- Data Entry Clerks
- Reconcilement Clerks

Key Steps

When establishing a Manual Testing Project there are several key steps that must be taken to order to facilitate the testing, including:

- Develop Test Process Flows
- Develop Input Documentation
- Develop Test Data
- Define Predicted Results
- Develop Test Verification Document Standards/Requirements

---

[45] Bret Pettichord, Quality Week White Paper, San Francisco, May 1996.

[46] Job Descriptions are included in Appendix I.

- Develop Manual Summaries and MIS
- Train the Test Team

Advantages/Disadvantages of Manual Testing

ADVANTAGES	DISADVANTAGES
People can be quickly redirected to new Tasks or tests	People get tired, make mistakes and can leave unexpectedly
Capacity can usually be expanded rapidly by using overtime, weekends and adding staff	People are expensive and require additional office support
	Excessive overtime and weekend work leads to diminishing returns

Figure 126 – Manual Vs. Automated Testing

Automated Testing

Automated testing can be used to improve the overall process of testing, including:

Authors' note: Test automation is a software Development activity and needs to be managed as such.

- Consistency
- Reliability
- Accuracy
- Efficiency

However, automation is not a solution to every testing problem, nor is it a guaranteed success. Most of the tools that are available are really software Development kits that can be used to build an automated testing Program. These Programs must be built using a SDLC and must be customized to each company's requirements. This information in this chapter and in the next chapter provides a basis for the customization.

Potential Automation Areas

Automation can be used to create Test Cases. Most testing professionals prefer to automate their Test Cases to produce consistent Test Transactions that can be repeated with minimal human intervention. Automation reduces the amount of time and resources that are required, and the Test Manager has the opportunity for broader test coverage. Automation also can support testing activities in a number of other areas including Stress Testing, Volume Testing, Code-coverage Testing, etc.

Once the Test Cases are available, there are different areas that are candidates for automated testing:

- The easiest step in automation is to replace the data entry people with some form of automated input. These people can be replaced by using scripting tools, keystroke-capture tools or a combination of the two. Once tests are input electronically, the management process can also be automated. This process can be limited to managing the entry of data, and in a more expanded form can be used to manage the remaining levels of test automation.

- The next step in test automation is usually to automate the comparison process, where the Actual Results are compared to the Expected Results.

- Automation can also be used to monitor the process of managing Defects and the progress of retesting.
- Many of the Systems Technical Tests that require repetitive activities or measurements can also be automated.
- And, in addition to automating the functionally oriented tests, automation also can be used to support some of the Systems and Business Technical Tests.

Staffing Requirements [47]

With automation, managers need fewer, but more expensive and more skilled people. The types of people who will be needed are:

- Technical Test Section Manager
- Tools Implementer
- Tools Administrator

Key Steps

When establishing an automated testing Project there are several areas that are specific to automated testing that must be considered, including:

- Establishing a flow chart of the complete testing process
- Determining which tests can be automated
- Selecting automation tools
- Developing the automated scripts
- Integrating the automated tools with:
  - Other existing Development / Management tools
  - The Application to be tested
  - Other Applications that will communicate with the Application being tested
- Defining specific MIS for the automated tools
- Training in the use of the tools

Factors Affecting Automated Testing

There are several factors that can influence the effectiveness of automated testing. The following factors should be considered when evaluating the need for a tool.

FACTOR	COMMENTS
Scope	The Scope of automated testing is based upon the dimensions of the Project: • Number of Test Procedures to run • Number of Modules/Functions to test • Number of Reports • Number of Interfaces • Number of Testers • Time frame for testing Tests should only be automated for a large volume of transactions and the Test Manager should not try to automate every transaction.
Preparation Timeframe	Automated testing requires more lead-time than manual testing. The tools must be

---

[47] Job Descriptions are included in Appendix I.

	selected, the staff Trained, and the tools Implemented.
	The time required to prepare to use automated tools should be built into the Test Plan timeline while the tools are being evaluated and selected.
Return on Investment (ROI)	The greatest gains with automated tools are achieved when the same tests will be performed more than once, either during the testing before the Application goes live, or during Regression Testing of Changes.
Test Integrity	There is a difference between a transaction that was successfully entered (i.e., passed the input edits) and a test that passed all of the processing and reporting logic that it was meant to test.
	Automated input tools must be used with some form of automated comparison tools to ensure that the Expected Result was obtained, or that the Defect is recorded.
Test Tracking	It is important to track the success (and failure) of transaction edit tests and logic tests. The Tool Implementer has to know what was not accepted and processed by an Application in order to know what to fix.
Installing the tool	Customizing an automated testing tool for a specific Application is like Developing a small Program. It has to be defined, programmed and tested.
Record and Playback	Record and Playback is useful only as a starting point. When an automated tool has this feature, a Tools Implementer can use this to capture a set of keystrokes; however, it is almost always necessary for the Developer to make adjustments to the automation script.
Test Independence	As transactions are entered automatically, the tools must be able to by-pass a transaction that does not pass the Application's edits so that the rest of the transactions can be entered.
	Transactions that are by-passed must be recorded so they can be repaired and re-entered.
Maintenance of Procedures	Most managers who decide to Implement automated testing tools intend to continue to use the tools after the Application moves into Production. This requires that the managers spend the time necessary to Document the process during testing and then to keep the set of Test Transactions current.

Figure 127 – Factors Affecting Automated Testing

Benefits of Automated Testing

The benefits of automated testing are related to the testing need, which is based upon the complexity and the size of the testing Project. There are different benefits that can be obtained from the different classes of tools.

Tool	Potential Benefit
Record / Playback	Record keystrokes and replay them
Code Analyzers	Monitor Code complexity, branches and dependencies
Coverage Analyzers	Cover multiple Conditions in the tests
	Cover multiple paths in the tests
Memory Analyzers	Implement bounds' checkers
	Install leak detectors
Load / Performance Test Tools	Monitor Application load levels
	Monitor web Application load levels.
Web Test Tools	Determine link validity
	Determine HTML Code usage validity
	Determine Client-side Program Functions
	Determine server-side Program Functions
	Evaluate web site's interaction security
Other Tools	Manage Test Cases
	Manage Documentation
	Manage Issues Tracking
	Configuration management

Figure 128 – Benefits of Automated Testing

Comparing Manual vs. Automated Testing

To identify which Test Cases are candidates for automation, the following factors should be considered:

- What is the fixed cost of automation?
- What is the variable effort in time and resources for each group of automation candidates?
- What is the manual effort that would be required to prepare, run and verify the same cases?
- Are there additional uses for the test tool and the Test Cases after this Project is completed?
- What is the long-term cost of leasing and maintaining the tool and the Test Library versus the expected use of the tool?
- Can the same tool be used to support other Applications?

Tools Supporting Testing During the SDLC

The tools available to support testing throughout the SDLC fall into several categories, as shown by the following chart.

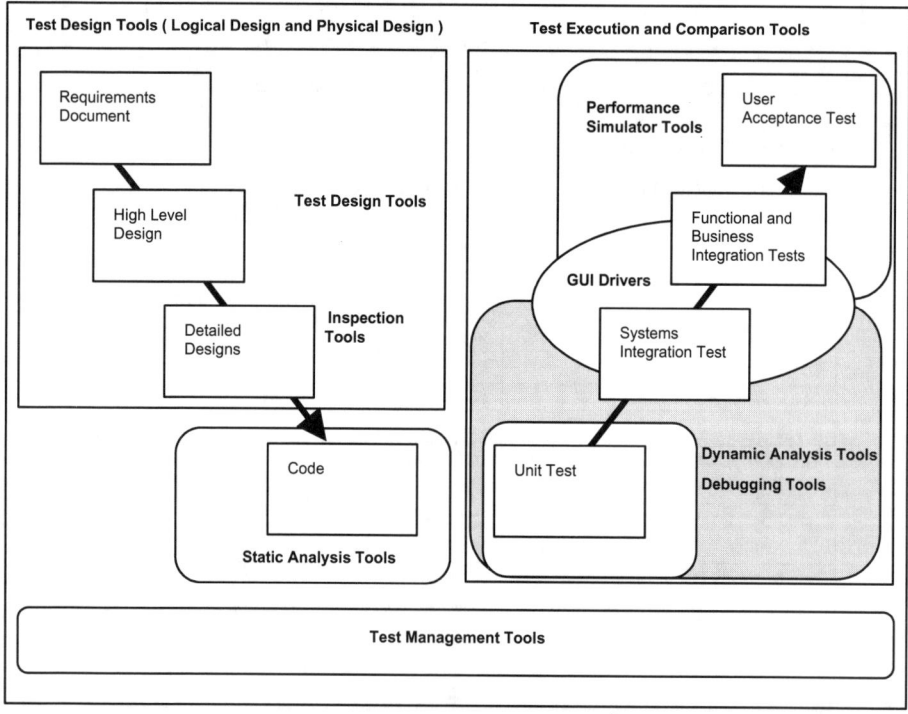

Figure 129 – Tools Supporting the SDLC [48]

This chart shows how the Development process evolves from the general Business Requirements to specific lines of Code, while testing builds from detailed Unit Testing to comprehensive User Acceptance Testing. And, within the chart, the

---

[48] M. Fewster and D. Graham, Software Test Automation, Addison-Wesley, 1999

categories of automated tools that are available to support the Code Inspections and Test Types are shown, and defined in the following sections.

Test Design Tools (Logical Design and Physical Design)

Test Design Tools

Tools that help the Test Manager decide what tests need to be executed.

Inspection Tools

There are three parts to automated software inspection: tools, flexibility and integration. Inspection tools are most effective in a distributed environment and are usually used via a network and/or the Internet. Generally, these tools facilitate collaboration more than automating the inspection process.

Static Analysis Tools

Static Analysis Tools analyze programs without executing the program. They focuses on descriptions, specifications and representations of software systems and should be performed before any other testing. Static analysis is used to:

- Check completeness and consistency
- Validate pre-defined rules
- Compare program logic with the specifications and documentation

Static testing can be performed manually or can be supported by tools, such as:

- Parsers
- Data flow analyzers
- Syntax analyzers

Test Execution and Comparison Tools

Dynamic Analysis Tools

Dynamic Analysis requires running the system to evaluate the Application before, during and after its execution in an artificial or real environment under controlled circumstances and with pre-defined Expected Results.

TYPE OF DYNAMIC ANALYZER	FUNCTIONALITY OF TOOL
Test coverage analysis	Tests to which extent the code can be checked by glass box techniques
Tracing	Follows all paths used during program execution and provides e.g. values for all variables etc.
Tuning	Measures resources used during program execution
Simulator	Simulates parts of systems, if e.g. the actual code or hardware are not available
Assertion checking	Tests whether certain conditions are given in complex logical constructs

Figure 130 - Dynamic Analysis Tools [49]

Debugging Tools

We have identified three categories of debugging tools.

---

[49] EAGLES Evaluation of Natural Language Processing Systems, EAG-EWG-PR.2, September 1995

Runtime Debugger

Code compilers almost always include a runtime debugger that is active as the Application is compiled. A runtime debugger allows Developers to compile and run with a single compilation, rather than modifying the source and re-compiling as they try to identify down the error. Although Runtime debuggers make it easier to detect errors in the program, they do not identify the cause of the errors.

Automatic Debuggers – Library Functions

Memory corruption and memory leaks can be detected automatically, and the Developer can be informed of the error. The simplest automatic debuggers have of a library of functions that can be inserted into a program. When the program is executed, these functions are called and the debugger checks for memory corruption; however, it cannot detect the point in the program where the memory corruption actually occurs.

Automatic Debuggers – On-Chip Instrumentation (OCI)

These tools read the object code generated by compilers, and before programs are linked they look for processor instructions that access memory, and can detect errors in dynamic memory, but they do not work on static memory and they cannot detect when memory leaks occur.

Automatic Debuggers - Scalable Coherent Interface (SCI)

The third group of runtime debuggers is based on SCI technology. The tool reads the source code of the program, and analyzes it to identify errors related to memory including heap, static, stack memory, language-specific errors and algorithmic errors.

GUI Drivers

GUI Test Drivers and Replay Tools automate the execution of tests for products with graphical user interfaces. This includes Client/server test automation tools and load testers.

Performance Simulator Tools

These tools specialize in putting a heavy transaction load on systems.

Test Management Tools

In some ways, all of the tools assist in test management. The following categories are specifically selected to assist the Test Manager in managing the testing effort.

Test Evaluation Tools

These tools that help Test Managers evaluate the quality of the tests.

Defect Tracking Tools

These tools that help Test Managers manage a database of defect reports.

Non-GUI Test Drivers and Test Managers

These tools automate execution of tests for products without graphical user interfaces.

Building Maintainable Tests

One of the primary advantages of automation is the Testers' ability to repeat tests with a minimum of effort. In addition to the tools that are required, there are two libraries that have to be maintained:

Test Libraries

Test Libraries contain the Test Cases, the Test Transactions and the Expected Results that can be used to validate the Application's Functions, and which can be reused with the same Application in the future or other Applications.

Test Suites

A Test Suite is a set of automation tools and the common rules that are needed for these testing tools to work within the Test Environment. These rules identify how the tools are to be used, not how the tests are to be constructed. Since interface technologies are constantly changing, a Test Suite can be used to make this process efficient.

Tool Vendors

There are several vendors that have built automated tools to test Enterprise Applications and Web-Based Applications, including:

VENDOR	ENTERPRISE	WEB-BASED
AutoTester	X	X
Compuware	X	X
Empirix		X
Mercury Interactive	X	X
McCabe	X	
Radview	X	X
Rational	X	X
Segue		X

Figure 131 – Vendor Coverage

A general description of the types of automated tools provided by these vendors is included in the rest of this chapter and each vendor's tools are evaluated in Chapter XVII (Automated Test Vendors).

Automated Tool Functions

There is a wide range of automation tools that support many different testing needs. The functions that are described in the following section are used in Chapter XVII to compare many of the different tools that are available from vendors.

Input

Some testing tools are available that automate the process of physically inputting data into the Application being tested.

Validation

There are two types of validation:

Input Edits

In this case, the testing tool automates the validation of input editing by determining if the transaction was accepted or rejected by the input Screen.

Result Validation

In the next level of testing the testing tool automates the result by holding the Expected Result in a File and compares the Actual Result that is obtained to that value.

Tracking

The testing tool can track the results of each element of the individual tests that are run, and Reports the results.

Scheduler

This type of testing tool can provide automatic Execution of predefined tests at predetermined intervals, or based on real time transaction activity.

Detail/Summary Results

In this case, the testing tool presents results of the data points that are tracked for each element of the individual tests that are performed and generates various Reports.

Process Management

A testing tool can provide Development and Testing Teams with an automated framework for managing the entire testing process from Design through Execution, analysis and reporting. The tool can help the Test Team:

- Plan tests easily
- Execute tests
- Vary the Test Environment
- Dynamically analyze the Application under test
- Track/analyze the Test Results
- Submit the Defects

Service Level Management

Based on predefined Application quality Service Level Requirements, the testing tool provides real time Application monitoring, diagnosis/cause identification, and alerts to technology managers.

Predefined Application quality Service Level Requirements typically include:

- Identification of the Application's performance bottlenecks
- System software/hardware infrastructure capacity problems
- Transaction response time from the End Users' perspective

In addition, monitoring these and other Production performance metrics can be used to forecast/Plan for efficient Application growth management.

Tool Evaluation

In order to evaluate a test automation tool for possible use, the Test Manager must first define what it is that the tool should do. Some examples [50] are:

- Consistent repeatable testing
- Run tests unattended
- Run tests more often
- Create better quality software
- Increase confidence in software quality
- Test more efficiently
- Test more software
- Increase user confidence
- Measure performance
- Reduce cost of testing
- Find more Defects
- Test on different Applications
- Earlier time to market
- Run tests more quickly
- Improve morale among Testers
- Test on different Databases

And, it is useful to define the criteria that will be used to compare and evaluate the available tools. Some of these criteria could be:

- Will the tool be used once for this Application, or reused over time?
- Is it important to save time now, or reduce the amount of time needed for testing over time?
- Will the tool be used in a single Environment, or multiple Environments?
- How many Testers will be using the tool?
- How sophisticated/experienced are the Testers?
- Will there be a single tool, or a set of integrated tools?

Tool Comparison Criteria

There are many criteria to consider when comparing tools, and the weight that a firm places on each criterion can be different. We have found it useful to do a rank score analysis when comparing tools. This process requires a firm to assign a weight to each question, and then to rate each tool on a 1 - 4 (or 1 - 5) scale to identify how well it meets each criterion.

The following chart lists some of the questions that firms should consider, and Test Managers should feel free to add their own questions. When assigning a weight, some firms prefer to allocate a total of 100 (or 1,000) points among the various criteria, thereby forcing them to discuss the allocation in detail. Other firms do not restrict themselves to a specific total, preferring to allow multiple

---

[50] ibid

people to put in a weight, and then calculate an average. Both approaches have worked successfully for many of our Clients.

The following example uses an arbitrary total weighting of 100 and a 1 to 4 relative rating. The weights assigned in this example to individual criteria are random. The Test Manager or a group of people can assign the weights. The result in the following example shows that Tool 1 is the better overall solution.

CRITERIA		WEIGHT	TOOL 1	TOOL 2	TOOL 1 WEIGHTED	TOOL 2 WEIGHTED
Overview	Setup	12	9	8	32	30
Product Features	Language Features	19	15	16	48	48
Transaction Test Process	Initiation Process	16	11	9	29	27
Load and Stress Test Process	Set-up Features	7	10	9	17	216
Performance Monitor	Product Features	5	5	6	14	14
Customer Support	Updates	14	13	6	44	23
Cost	License Cost	4	4	2	16	8
Database Requirements	Database Features	2	4	4	8	8
Interfaces	Applications	9	10	9	25	20
Technical Support	Support for Multiple Platforms	8	7	7	16	18
Other	Product Comparison	4	3	3	6	6
	TOTAL	100			255	218

Figure 132 – Sample Tool Comparison Matrix [51]

Technical Considerations

When comparing automated tools, the checklist previously provided is one approach; another approach that can be used with the checklist or alone is to evaluate the technology behind the tools. Some considerations are:

Testing Methodology

When a vendor provides metrics for their tools, the Test Manager should ask about the number of machines, locations, or IP addresses that were utilized in to establish the measurements. KeyLabs[52] has shown that there is a point at which any load generating tool stops generating an accurate load. In fact, depending on the complexity of the website, this point of diminishing returns can become a factor with as few as 30 virtual users per machine.

Using Multiple Protocols

Some Applications require the use of multiple protocols, and therefore the testing effort must also consider that situation. For example, to test the loading of an http page, it may be necessary to log in to an Application using a Java applet on that page, upload some data via FTP, update a Database via an http form, and

---

[51] The details behind this Scorecard are available from the Authors.

[52] KeyLabs provides the testing industry with a full suite of network testing services that includes e-commerce stress and security testing, SAN testing, performance analysis, scalability analysis, and proof-of-concept testing. www.keylabs.com

finally verify the activity by interacting with the Database itself to determine that the data arrived properly.

This test involves http, FTP, Java and JDBC. For each of the four protocols, some vendors must generate four distinct scripts to make this work – other tools may only require that the script be written once. This load must be considered when estimating with a tool.

Executable

Users must evaluate whether compiled or Executable Code provides the best support for a given situation. Some tools also require that a separate copy be run for each virtual user, others do not.

Pricing Model

With some tools a user must purchase a controller as well as multiple user licenses. Some tools require a separate license for each protocol.

Java Components

Some tools use the Client's native Java Environment while other tools replace the Client's JVM (Java Virtual Machine) with their own JVM for performance reasons.

IP Spoofing

IP Spoofing may allow users to run multiple different IP addresses from a single machine and use all of the IP addresses in a load test. When KeyLabs attempted to use this feature in a test, it did not work for all tools.

General Performance

The general performance of the tool under a variety of situations is also evaluated. Some of the situations that are considered are:

- The testing tool takes a considerable amount of time to launch
- Data collection after a test is slow
- Data may be corrupted if the tool allocates all of the resources at the start of the test for the entire test level

Summary

The primary points covered in this chapter are:

- The difference between White Box Testing and Black Box Testing, where in White Box Testing the Testers and the End Users exercise the Functionality of the Applications without knowing exactly how the code works. In Black Box Testing, the Developers test the code by examining how the code works.
- The chapter presents the strengths and weaknesses of automated testing versus manual testing, and proposes a methodology to compare the two.
- Automating GUI testing requires some special considerations.
- Some of the leading automated testing tool vendors are identified and the types of tools available are discussed. A set of comparison criteria is also presented.

XIV. CLOSING THE TEST PROJECT

There are a number of different ways that a Project can come to a Close. The preferred way is when the Project runs through to completion, with all of the stakeholders agreeing that the Application that is delivered meets the original/updated objectives. The least desirable way is a premature conclusion that results from business strategy changes, changing Client needs, a withdrawal of funding, or an inability to get the Application built and delivered on time.

Although many of the Closing Tasks would be the same for any of these situations, this chapter reviews the activities that can be used with a successful Project. This could include Projects that:

- Meet all of the originally stated objectives
- Projects where, for any number of reasons, the deliverables were sub-divided so that they could be installed in multiple phases

There are two distinct sets of activities that must be performed to successfully Close a Project.

- The first set of activities consists of the steps that must be taken prior to Implementation to ensure that all of the Project milestones have been reached. If all of the milestones are successfully completed, the Project Manager can be more comfortable moving the software into Production.
- The second set consists of those Tasks that should be performed after the Application has been Implemented to ensure that the Project is appropriately concluded from an administrative, people and process point of view, and to prepare for the:
  - Future changes to this Application that will require Regression Testing
  - Future use of the Test Cases and Test Procedures for other Applications

Both of these sets of activities are covered in this chapter.

Prior to Implementation

Prior to Implementation, all of the stakeholders must agree that the key milestones have been successfully met, and that everyone is comfortable with moving the Application into Production. To prepare for this, the following questions must be answered:

- Is the Application ready?
- Has all of the required information been successfully converted?
- Is the technical Environment ready?
- Are the End Users ready to respond to the new business?
- Are the End Users prepared to use the Application and any new procedures?
- If Clients are going to use the Application, are they ready to use it?

Although it is primarily the Project Manager's responsibility to ensure that all of these questions get answered affirmatively, it is the Test Manager (with input from the entire Test Team and the Project Management Team) who typically provides the bulk of the feedback on whether the software is ready to move into Production. For the Test Manager, this final step is really an extension of the evaluation process that has been conducted throughout the entire Project.

As the Project nears completion, the Test Manager must know the Status of any:

- Core Functions that have not been successfully tested
- Defects that are outstanding

A review of these two factors ultimately determines whether the software itself is ready for Production, and the Test Manager should be prepared to review this information with the Project Management Team, and in some instances, with senior management.

Identify Software to Move into Production

Decisions are normally made throughout the Project's life cycle to defer some Functions and Defect repairs until after the initial Implementation. In most large Projects, there is a last-minute debate concerning which Functions will be installed in the first Implementation. In order to support this decision, the Test Manager should have a summary of the:

> **Authors' note:** The decision to move an Application into Production is a joint decision. It is normal in large Projects to defer some functionality until after the initial Implementation.

- Most recent status of the Functions that have been successfully tested and ready for a move to Production
- Functions that will not be ready for Production
- Current status of the Defect repairs

Any of the Documents that were prepared earlier in the Project that contain a list of the Planned Functions could be used as the basis for a review of the status of each Function, and the list of the outstanding Defects should be added. The following is a sample format that could be used for this evaluation.

SOFTWARE MOVE CONTROL LOG					
TYPE		FUNCTIONS	MOVE TO PRODUCTION	DEFER	COMMENTS
Functionality	Screens	Logon	X		
		Trade Entry	X		
		Settlement	X		
		Posting	X		
		Entitlements		X	Must be in by the end of the 1st Qtr.
	Interfaces	Cash Movement	X		
		DTC	X		
		Reconciliation	X		
		General Ledger		X	Must be in by 1/1
	Reports	Daily Activity	X		
		Failed Trade	X		
		Position Reconciliation	X		
		Annual Reports		X	Must be in by 12/31
		FASB		X	Must be in by 12/31
Defects	Issue 1		X		

	Issue 2		X	
	Issue 3		X	Manual workaround in place. Automation must be in by 1/1
	Issue 4		X	Must be in by 12/31 to support new Client base.
	Issue 5	X		
	Issue 6		X	Manual workaround in place. Must be in by 12/1 to handle anticipated volume increase.

Figure 133 – Sample Software Move Control Log

Deferred Functions

The Test Manager must ensure that there is a solid Plan in place to accommodate the future testing of any deferred Functions. If a manual workaround is available as the solution, procedures should be documented, and the process tested with the procedures in place. Similarly, if the solution consists of a combination of a partially automated solution in conjunction with a manual workaround, both parts of the process should be tested.

Authors' note: One of the Risks with a workaround is that the software Change that is needed cannot be cost justified on its own after the Application goes live.

For example, if a particular Transaction Type is not working, it might be possible to use two existing Transaction Types to obtain the same accounting result and link the transactions by temporarily modifying the description fields. This process must be tested to ensure that it can create the desired results until the ultimate Changes are made. It is important for everyone to understand the workaround process, the results, and how the end result will appear to the users and Clients of the Application.

Sign-offs

Ideally, sign-offs will have been obtained as each Test Phase was completed. Typically, this includes sign-offs on specific Functions, along with approvals to move from one Test Day to the next, and one Test Phase to the next. The final set of signatures should represent the approval to move the agreed upon software into Production, along with an agreement (if relevant) that specified Components are deferred until a later time.

Authors' note: The Test Manager must obtain formal sign-offs for the testing activities throughout the Project life cycle.

There are three general ways to obtain sign-offs:

- Obtain sign-offs as a routine part of the Test Process by adding signature lines to the Software Move Control Log
- Require that each of the stakeholders send a note to the Test Manager to indicate their approval for moving forward
- Hold a sign-off Meeting to obtain all of the signatures at one time

Each of these three methods has benefits and shortcomings, but the authors prefer the first alternative.

Post Implementation

After the software has been moved into Production, several wrap-up activities must occur. These include the following:

- Developing Test Plans for the ongoing activity (if applicable)
- Rewarding/recognizing the people who made the Implementation a success
- Closing the Financials
- Conducting a Post Project Review to learn from successes and mistakes

Each of these is discussed in the following sections.

Developing Test Plans for Ongoing Activity

Testing the remaining Functions must be Planned, Executed and Validated. Ideally these Plans would be completed before the Application moves into Production; however, as is often the case with a large/complex Implementation, the Test Manager and Test Teams focus almost exclusively on any last minute testing that is absolutely required for the Implementation.

If detailed Plans for the future were not completed previously, the Test Manager must create them immediately after Implementation. This Plan should include:

- Staffing Requirements
- Budget Forecast
- New Test Calendar
- Redeployment Plan for any of the Test Team members who are not needed on the remaining activities
- Termination or redeployment of consultants

If it is necessary to terminate either an individual consultant, a group of consultants from one agency, or to terminate an entire outsourcing arrangement, the Test Manager must either follow the:

- Rules established in the relevant contract
- Rules established by the organization
- Industry accepted Guidelines if one of the first two points does not exist

> **Authors' note:** The industry's termination Guideline is a two-week notice.

As one of the final steps in this process, the Test Manager must also make a decision about what to do with the vast repository of information that likely has accumulated.

- If the Project is ongoing, the Test Manager should review the future Plans and decide what should be archived, and what should be actively kept open.
- If the Project is truly coming to a Close, the Test Manager should decide what Documentation should be discarded and what should be archived. This archiving could include the full range of test materials, including Test Procedures, Expected Results, etc., or a sub-set of the Documents.

> **Authors' note:** Save your work! Whenever the authors have not done so, we have regretted it when we started our next similar Project.

At this time in the Project to the Test Manager should archive the Business and Functional Requirement Documents along with all of the related notes, and to clear the closed Issues in the Issues Tracking Report, leaving only the open Issues active. Additionally, it often makes sense to establish a Test Library if one does not already exist, and populate it with the Test Cases and Test Procedures that were written for this Application so that they can be used for future Regression Tests.

It is up to the Test Manager to determine what should remain open, and what should be archived, and when. However, when the Project ends the key Documents and Databases should be saved, either as reference material for future Projects and for use by the Audit department, or as a foundation upon which to build another set of Plans, etc.

> "The further backward you look, the further ahead you can see."
> — Winston Churchill

Rewarding/Recognizing People

Rewarding and recognizing people are often overlooked steps in the Closing process, and unfortunately they are often ignored throughout the entire Project life cycle. The people who make an Implementation a success should be rewarded commensurate with the result and their overall contribution. This does not (and cannot) always translate into a financial reward; however there are many other ways to show appreciation for the long hours, days and months of effort.

Although these things are not always within the Test Manager's control, the following list provides some suggestions for the Test Manager, when and where appropriate:

- Personal Recognition
  - Words of encouragement, thanks and appreciation (the cheapest of all, but very well received!)
  - Celebratory events: breakfasts, lunches, dinners, parties
  - Tokens of appreciation: pens, mugs, hats, jackets, etc.
  - For significant contributors: Nights on the Town, Dinners, Expense-paid mini-vacations, etc.

> "Leadership is the art of getting someone else to do something you want done because he wants to do it."
> — Dwight Eisenhower

- Career Enhancing Recognition
  - Provide widespread exposure within the organization
    - Congratulatory memos
    - Newsletters
  - Provide opportunities for:
    - Job expansion
    - Promotion
    - Newer/diverse experiences
- Financial Recognition
  - Bonuses
  - Salary increases

People contribute to the success of a Project in varying degrees, and as a result, not all of the reward/recognition items are appropriate for every person on the team. It is up to the Test Manager to evaluate each Test Team member's contribution to the Project, and make every effort to reward accordingly.

One of the least costly and easiest things to do is to express appreciation and gratitude to each individual contributor as appropriate for their efforts. This should not only be done at the conclusion of the Project, but throughout the life cycle of the entire Project. Additionally, it is important to let people know how important their contribution is/has been to the success of the Project.

> **Authors' note:** The absence of praise when it is deserved is a de-motivator for most people.

Other rewards can cost money, and the Test Manager will need to honestly assess who put forth the biggest effort and/or made the most significant contribution(s). It is this group that usually gets some kind of financial reward, along with possible career enhancing opportunities. The Test Manager may have to work hard to get an organization to recognize and support a financial reward for members of the Test Team. In some organizations, but not all, the Test Team members are considered less important than the Developers and the End Users, and as a result, as money is being appropriated across the entire Project Team, testing people often fall to the bottom of the list. If this situation exists, the Test Manager should be realistic in the submission of names, while at the same time making sure that he/she does everything they can to get the best people appropriately compensated.

Closing the Financials

It is very tempting to stop updating the Financial Reports once an Application is Implemented. However, the Test Manager should invest the time to bring the Financial Reports up-to-date, not only for his/her own benefit, but also for the organization's use. Hopefully, throughout the entire Project, the financials were properly maintained, and this last step should include the final bills for the Project after they arrive.

For the Test Manager, it is important to take the time to evaluate where there were variances to the Plan, and understand which variances were under his/her Control. Understanding these variances can help guide the Planning for any future Projects.

It could be that the Test Manager totally underestimated the cost of the consultants, or how much time a particular Test Phase would take. This information prepares the Test Manager to make more realistic projections in the future.

Additionally, the Project Manager should consolidate the testing financials with the other Project expenses, and this consolidated Report will provide the Project Management Team with a complete picture of how the entire Project did compared to the original estimates. This information is useful not only for future Planning but also for making potential adjustments to the current (and possibly future) business Plans/Budgets to compensate for any variances.

> **Authors Note:** Any negative surprise is unpleasant. Negative financial surprises are disastrous for Projects and careers.

In a well-managed Project, the financials for the entire Project will be kept up to date with periodic updates given to the senior managers. The Test Manager should keep the testing financials current, and force a periodic review even if the Project Manager, and/or senior managers do not require it.

If the financials have been maintained, there should be few surprises when the final Report is completed. However, if the financials were not maintained, and this information is being assembled and communicated for the first time, it could be a surprise. If the first time that the senior managers see and/or hear of significant negative variances is at the end of the Project, it could result in unpleasant repercussions for the Project Manager, as well as the Test Manager.

Post Project Review

The lessons learned throughout a Project can be invaluable if they are captured, understood, and acted upon when appropriate. It is wise to gather this information immediately after an Implementation, while the circumstances are still fresh in everyone's mind, to identify what did not work, what was a clear success, and where there was room for improvement. This process should involve everyone involved with the Project, but will also be worthwhile if it is only used within the Test Team. Regardless of how it is done, there are a few basic steps that should be followed, including:

- Solicit/gather the feedback (non-judgmentally)
- Review/assess the feedback
- Document/incorporate/communicate meaningful activities based on the feedback

Although it often the last thing a Test Manager wants to do after a complex Project, it really is in the Test Manager's best interest to participate in the overall process, or capture this information from the Test Team. If managed properly, without assigning blame, the review can identify shortcomings in the process and highlight practices that worked particularly well, which in the long run can only benefit a professional manager.

> "A real leader faces the music, even if he doesn't like the tune."
> Anonymous

Solicit / Gather Feedback

One of the first things the Test Manager must decide is who should participate in the feedback process.

- Often the process is limited to a few key people in the hopes of capturing the most pressing issues without involving a large group of people. The benefit to this approach is that it usually can be done very informally and quickly.
- Another approach is to solicit feedback from every member of the Project Team or Test Team. Although this approach will result in a lot of information and a lot of emotion, the best recommendations often come from the most unexpected sources. In limiting the participants, the Test Manager runs the risk of missing out on some very helpful suggestions.

Regardless of the approach, a process must be established to gather the information. The most common methods include:

Informal Method

The Test Manager could hold a Meeting where each participant is asked to come to the Meeting prepared to provide feedback. The feedback should be documented and incorporated into minutes of the Meeting.

Formal Method

There are two formal methods:

- In a structured process, the Test Manager should prepare a form to be completed by the participants, and subsequently reviewed in a Meeting. If this method is selected, the Test Manager should decide on the format and content of the form, which could look like the following example:

\multicolumn{2}{c}{TESTING POST PROJECT REVIEW}	
Project: _____	Submitted By: _____
	Date: _____
Phase: _____	Reviewed By: _____
	Date: _____
Form Number: _____	
TOPIC	**SUBMITTER'S COMMENTS/SUGGESTIONS**
Planning	
Execution/Verification	
Meetings	
Tools	
Other	
Reviewer's Comments/ Actions	

Figure 134 – Sample Testing Post Project Review Form

- A less structured process is to solicit information in pure narrative form. Sometimes this approach yields less linear thinking, as people do not follow the thought process established by the form.

In any case, the most important thing is to solicit the feedback, document it, and take it seriously as noted below.

Review/Assess Feedback

Although the testing effort is clearly a group effort, it is usually the Test Manager who defines the process, and oversees the Planning, Execution, Validation and Communications. And, it is normally the Test Manager, possibly along with a few direct reports, who will evaluate the feedback, and make decisions around the changes that should be made.

> **Authors' note:** "Judge the future by the past."
> Patrick Henry

If the Post Project Review is conducted on the whole Project, the Project Manager will certainly be involved in examining the testing problems and suggested solutions. The Test Manager should not take the feedback/criticisms personally, and should evaluate the information with an open mind. Often it is the Test Manager's direct reports and/or the other reviewers who help the Test Manager see something that might have been overlooked due to personal biases.

Document/ Communicate/ Incorporate Relevant Adjustments

Once the participants agree upon the adjustments to recommend, they should be documented, communicated to the participants as well as to senior management if they were aware of the Issues, and integrated into the formal testing process as appropriate. One way to document and communicate the adjustments is to consolidate the information gathered, produce a comprehensive summary of the information including the proposed adjustments, and re-group the original participants for an update. This process allows the interested parties to challenge why other adjustments are not going to be made, and provides the forum for the Test Manager and/or review group to substantiate their reasons.

This process is usually performed and is the most successful when the participants are part of a relatively static team that will continue as a group on other near-term Projects. However; if the Project Team will be disbanded when the Project is completed, it is still a good idea for the Test Manager to document the findings, so that they can be incorporated into future Projects either within the same group or even outside of the group. The lessons learned are usually universal, and by documenting them and making them available for review could help the organization avoid repeating some of the potentially costly mistakes encountered in the current Project.

Summary

Properly Closing a Project benefits all of the parties involved. A professional Closing process recognizes the people who made the Project successful, it retrieves a wealth of information from those very same talented people regarding what worked and what did not work throughout the Project, it forces the Test Manager to go through a diagnostic process that should only make him/her a better manager, and it provides the organization with invaluable information for future Projects.

> **Authors' note:** Much of this book is the result of lessons learned during intensive Post Project Reviews.

There are two key things that must take place to bring a Project to a successful Close, which are described in the following sections.

Pre-Implementation

The Project Manager must ensure that all of the agreed upon Project milestones and the Pre-Implementation milestones, including the following, have been met:

- Application Software is ready
- Technical Environment is ready
- End Users are ready to:
    - Handle the new business (if applicable)
    - Use the new Application and Procedures
- Clients are ready to use the new software, procedures, reports, etc.

Post Implementation

The Post Implementation review should ensure that the Project is appropriately concluded from the following viewpoints:

- Administrative
- People
- Process

The Post Implementation milestones include:

- Developing Test Plans for Ongoing Activity (when relevant)
- Rewarding/Recognizing People
- Closing the Financials
- Conducting a Post Project Review

Aside from the process being time consuming, and potentially resulting in some bruised egos, there is very little downside to performing these activities. In the authors' experience the benefits gained from the individual players, as well as the organization, definitely make it worth the effort.

XV. AUTOMATED TEST VENDORS

Several automated tool vendors have been evaluated and are presented in alphabetical order.

AutoTester

AutoTester ONE

AutoTester ONE provides an end-to-end testing solution with Functional, Regression, and Business Integration testing for Windows, Client Server, Host/Legacy, ERP or Web Applications. In addition, the AutoController product provides stress and load testing for Applications on Windows, ERP, and the Web.

Product Features

The main product features are:

- Captures tests as a series of objects and data-aware events that populate a highly flexible, reusable and maintainable test library
- Adjusts during playback to accommodate Changes in size or position of Application icons, objects, or controls
- Provides verification of the Application objects' contents and states, displays text values, and complex bitmap images during test playback
- Allows for unexpected Application or hardware responses during Execution, and adjusts in order to continue testing
- Logs the results of Application failures and Defects and then continues the testing process
- Supports playback synchronization
- Tests across all types of Applications with the same Test Procedures, and Functional Test Procedures can be used for Stress Testing
- Includes Test Management capabilities and a Test Scheduler for unattended testing
- Does not require programming experience to modify and maintain scripts
- Produces self-documenting Test Procedures and instant Test Results

AutoTester believes that AutoTester ONE is the only automated testing product that integrates Windows Desktop, Web, Host/Legacy and Client Server testing capabilities, including a fully integrated Test Management Module.

Platforms

AutoTester ONE is Client-resident and supports any Applications that can be accessed through all Windows Operating Systems, regardless of platform.

AutoController

AutoController is a Stress Testing tool that utilizes the Test Procedures written for the tests that are run by AutoTester ONE to provide end-to-end Stress Testing of Applications. This tool can support Stress Tests for Windows Desktop, Web, Host/Legacy and Client Server Applications.

Product Evaluations

SDLC Support

AutoTester ONE supports all facets of the SDLC during Development and Maintenance testing.

	Development			Acceptance Testing				
	Unit	System	Functional	End-to-End	UAT	Load/ Stress	Performance/ Tuning	Business Integration Testing
Auto Tester ONE	X	X	X	X	X	X	X	X

Figure 135 – Support During Development Projects

	Development				Acceptance Testing				
	Unit	System	Functional	Regression	End-to-End	UAT	Load/ Stress	Performance/ Tuning	Business Integration Testing
Auto Tester ONE	X	X	X	X	X	X	X	X	X

Figure 136 – Support During Maintenance Projects

Product Features [53]

AutoTester ONE supports Enterprise and web-based Applications.

	Features					
	Enterprise Applications Tool	Web Applications Tool	Other Applications Support	Identifies Code Defects/ Run Time Defects	Identifies Untested Code	One Script for All Applications
Auto Tester ONE	X	X	Windows Desktop, ERP (SAP), Client Server, Mainframe/Host/Legacy, Citrix	X	X	X

Figure 137 – Product Features

Product Benefits

AutoTester ONE offers most of the basic benefits expected in a testing tool.

	Benefits						
	GUI Based Applications Test	Scripting Language	Script Development Modules	Predictive Capability	Change	Verification	Single Point of Control
Auto Tester ONE	X	X	X		X	X	X

Figure 138 – Product Benefits

	Functions						
	Input	Validation	Tracking	Scheduler	Detail/ Summary Results	Process Management	Service Level Management
Auto Tester ONE	X	X	X	X	X	X	

Figure 139 – Product Functions

---

[53] The definitions for the column headings for this chart, and subsequent charts in this Chapter, are included in Appendix III.

Compuware

Compuware license revenue for their Fiscal Q4 2001 was $138.2M, where $94.67M (68.5%) was from Mainframe Clients and $43.53M (31.5%) was from Distributed Software, of which $8.9M was from testing. Compuware's testing tool business is growing at the rate of approximately 23% annually.

QARun

QARun 4.8.0a supports the testing of Web, Java, ERP, CRM and internally Developed Applications. The tool provides the means to automate manual Tasks, create repeatable tests and provide for viewing, verifying and analyzing Test Results.

Product Features

The main product features are:

- Central repository
- Wide Support
- 24X7 Hotline
- GUI Interface
- Easy-to-read script
- Generic mapping
- Code coverage
- Dynamic Java control mapping
- Microsoft Access 2000 (supplied with QARun)
- SQL Server 6.5 and 7.0
- SQL Server 2000
- Oracle 7.3 and 8

QARun supports four distinct levels of Object Recognition:

- API – QARun gets the internal object name and extracts the required information
- Standard – Can use hWnds Window handles
- Derived – QARun can utilize a Standard Window that acts like a control, and which can be mapped manually to an alias
- Emulation - Learns positional mouse clicks

QALoad

QALoad performs repeatable load testing to determine the ultimate performance and potential limits of an Application. It helps process loads that mimic realistic business usage as well as validate that the Application can meet acceptable Service Levels.

Product Features

The main product features are:

- Accurate
- Unique cookies

- IP Spoofing
- Compiled ANSI C Code vs. interpreted languages which reduces the amount of RAM required on each testing PC
- Small footprint
- Dial-up/Dial-down
- Generic virtual users
- Unlimited analysis agents
- Unlimited Player platforms
- .NET support
- QALoad shows if software breaks, when it breaks and where it breaks
- Unique proxy-capture mechanism that allows users to capture information from any browser on any device.
- Accurate multi-threaded Unix playback mechanism.
- Monitoring is built into QALoad and QACenter Performance Edition.

TestPartner

TestPartner provides automated Functional Testing of complex Applications based on Microsoft, Java and Web based technologies. It allows both Developers and Testers to create repeatable tests through visual scripting and automatic wizards and also allows access to the full capabilities of Microsoft's Visual Basic for Applications (VBA).

Product Features

The main product features are:
- VBA scripting language
- Visual testing interface
- Dynamic ActiveX mapping
- 'Headless' testing of Communication objects

QADirector

QADirector gives Development and Testing Teams a framework for managing the entire testing process (from a single point of control); for the full life-cycle testing of distributed, large scale Applications from Analysis and Design through to Execution.

Product Features

The main product features are:
- Workflow
- Central Database
- Manage the test cycles and allow the scheduling of the tests and test suites on separate machines at different times of the day

DevPartner

DevPartner provides Application and Database Development productivity/reliability testing tools that help Developers automatically detect, diagnose and facilitate resolution of software Defects, and measure Code performance and Code coverage during testing. The tool supports simple two-tier Applications as well as complex distributed and web-based Applications.

Product Features

The main product features are:

- Visual Studio and VisualStudio.NET support
- Java Support
- IDE integration
- Distributed analysis
- Kernel level analysis and debugging
- Support for driver Development
- SQL and stored procedure analysis, debugging and tuning

PointForward

PointForward provides remote web performance testing and monitoring. It includes web Application scalability, integrity, and reliability testing capabilities.

Product Features

The main product features are:

- Remote Execution
- On-site expertise
- Execution from multiple points of presence
- Hands-off approach

Strobe

Strobe is a full life-cycle performance management solution for measuring and analyzing the activity of OS/390 and zOS online and batch Applications.

Product Features

The main product features are:

- Code-level performance analysis
- DB2 Support
- MQ Support

Vantage

Vantage is a full life-cycle performance management solution for distributed Applications that includes several products:

- Server Vantage
- Network Vantage
- FileAID/CS (also included in QA Center)

The Vantage tools monitor the Application from End User through the network to the server and Database tiers. Vantage identifies the precise nature and causes of performance bottlenecks and predicts future network and Application performance.

Product Features

The main product features are:

- Workstation, Network, Server and Database real-time performance analysis
- Alerts
- Network Prediction capabilities
- Application traffic analysis - ID apps accessed
- Determine Code bottlenecks
- Analyze network latency
- Determine or Predict Network performance

Product Evaluations

SDLC Support

	Development		Acceptance Testing					
	Unit	System	Functional	End-to-End	UAT	Load/Stress	Performance/Tuning	Business Integration Testing
QARun	X	X	X		X			X
QALoad				X		X	X	
TestPartner	X	X	X					X
QADirector	X	X	X	X				X
DevPartner	X		X					
PointForward				X		X	X	
Strobe							X	
Vantage				X		X	X	
Xpediter	X	X						

Figure 140 - Support During Development Projects

	Development		Acceptance Testing							
	Unit	System	Functional	Regression	End-to-End	UAT	Performance/Tuning	Load/Stress	Business Integration Testing	Scalability Testing
QARun	X	X	X						X	
QALoad					X		X	X		X
TestPartner	X	X	X	X		X			X	
QADirector	X	X	X	X			X		X	
DevPartner	X		X							
PointForward				X		X	X	X		X
Strobe							X	X		X
Vantage							X	X		X
Xpediter	X	X								

Figure 141 - Support During Maintenance Projects

Product Features

	Features					
	Enterprise Application Tools	Web Application Tools	Other Application Support	Identifies Code Defects, Run Time Defects	Identifies Untested Code	One Script for all Applications
QARun		X	eBus/ERP/ CRM Central object repository (ease of Maintenance), Object recognition, Capture/replay scripting, Application support, Dynamic Java Control support			X
QALoad	X	X	eCom/ERP/CRM Multi-Function vu's, true accuracy, performance statistics, EasyScript technology (universal scripting), unique cookies, IP spoofing			X
TestPartner	X	X	VBAScripting, Communication Object testing, "Headless" testing, Integrated internally with SAP, Dynamic ActiveX and Java Control support			X
QADirector	X	X	PDAs Controls testing procedures, mixed testing types, tracks Defects, manages and associates Requirements, management Reports, central Reporting repository		X* *Active analysis with DevPartner	
DevPartner	X	X	VB, VB.NET, C++, C#, ASP, Java, WebScript Code coverage, Code performance, Defect detection, memory profiling, memory leak detection	X	X	
DevPartnerDB			Oracle, Sybase, SQL Server Stored Procedure & SQL Debugging; SQL tuning for performance, Auto Tuning (SQL Server and Oracle)	X		
SoftICE			Windows Kernel-level debugging Stop-start, freeze, breakpoints	X		
Driver Studio	X	X	Windows Device Drivers Only Code coverage, Code performance, Defect detection, memory profiling, memory leak detection	X	X	
PointForward	X	X	Remote Execution; remote expertise, performance statistics			X
Strobe	X		OS/390, DB2, CICS Performance statistics down to the line of Code			
Vantage	X		Distributed Sys: UNIX and NT Servers, Networks Network, CPU and Database health 24X7; network performance and activity, Application performance down to line of Code			
Xpediter	X		OS/390, CICS, TSO, IMS, Cobol, Assembler, PL/I debugging	X	X	

Figure 142 – Product Features

Product Benefits

	Benefits				
	GUI Based Applications Test	Scripting Language	Script Development Modules	Test Creation	Predictive Capability
QARun	X	X	X	X	
QALoad	X	X	X	X	X
TestPartner	X	X	X	X	
QADirector	-	-	-		
DevPartner	X	-	-	X	
PointForward	X	X	X	X	X
Strobe	-	-	-	-	X
Vantage	-	-	X	-	X
Xpediter	-	-	X	-	X

Figure 143 – Product Benefits

Product Functions

	Functions						
	Input	Validation	Tracking	Scheduler	Detail/ Summary Results	Process Management	Service Level Management
QARun	X	X	X	X	X		
QALoad	X	X	X	X	X		X
Test Partner	X	X	X	X	X		
QADirector	X	X	X	X	X	X	
DevPartner	X	X	X	NA	X	X	
PointForward	X	X	X	X	X		X
Strobe				X	X		X
Vantage			X	X	X		X
Xpediter			X		X		

Figure 144 – Product Functions

Platforms

These Compuware products support the following platforms:

	WINDOWS	UNIX	LINUX
QARun	Yes	No	No
QALoad	Yes	Yes	Yes
TestPartner	Yes	No	No
QADirector	Yes	Yes	Yes
DevPartner	Yes	No	No
PointForward	Yes	Yes	Yes
Strobe	Yes	No	No
Vantage	Yes	Yes	Yes

Figure 145 – Platforms Supported by Compuware

Mercury Interactive

Mercury Interactive Corporation is an industry leader, with testing license revenue for Fiscal Q4 2001 of $61.2M, and a growth rate for testing tools of about 60% annually.

LoadRunner

LoadRunner is a load-testing tool that predicts Application behavior and performance. It exercises an entire enterprise infrastructure by emulating thousands of users and employs performance monitors to identify and isolate problems.

The primary feature is that LoadRunner uses C-based scripts, which run as an interpreted language.

XRunner

XRunner automates Functional testing to ensure X Windows–based Applications work as expected. It records business processes into Test Procedures, supports script updates as the Application is Developed or Changed, Executes scripts, Reports results and enables script reusability when an Application is modified or upgraded.

ActiveTune

ActiveTune is a service specifically designed for tuning Production Applications. It is designed for Applications that are already in Production, being upgraded, or about to be launched. ActiveTune allows internal experts to work with Mercury Interactive's tuning specialists in the areas of infrastructure, Application and security.

TestDirector

TestDirector is a complete, open-ended Test Management framework that enables management and Control of all phases of Application testing. TestDirector's Function can be extended to work with other testing tools and return results to the central repository.

Astra FastTrack

Astra FastTrack is a Web-based Defect management tool. It enables companies to Report, review and analyze Defects to determine Application readiness. It's 100% web enabled Interface allows any member of the testing team to access Defect information via a browser and prepare a variety of Defect Reports.

WinRunner

WinRunner is an integrated, Functional testing tool for the entire enterprise. It captures, verifies, and replays user interactions automatically to verify transactions and identify Defects. Test creation and debugging works through the use of a GUI, and a context-sensitive menu that allows identification of highlighted script objects within the Application.

Test Termination

WinRunner tests can be run in unattended batch mode until they terminate unless an explicit action is taken to stop the test early. This means that a test that has a fatal testing Defect could continue running until a person stops them.

When Executing tests in non-batch mode (Debug, Verify or Update modes), WinRunner will present a dialog box if fatal Defects occur.

Add On Programs and Extensions

The basic WinRunner product can work with objects and windows that were built using the standard Microsoft Foundation Class (MFC) library. Objects and windows Developed in any other way have to be treated as custom objects.

Object Hierarchy

WinRunner typically has a flat object hierarchy where child objects exist in parent windows.

Language Localization

WinRunner does not have any Documentation regarding how to use the product to test language-localized Applications.

Database Verification

WinRunner has some built-in capability to make SQL queries that can extract information from an ODBC compliant Database and save it to a File.

The Developer must Code any data verification.

Data Driven Testing

WinRunner supports the process of data-driven tests by using an Excel compatible spreadsheet File with data in the cells.

Recovery

WinRunner does not provide a built-in recovery Application.

Scripting Language

WinRunner has a proprietary, interpreted, scripting language that is a C-like procedural Programming language called TSL.

Exception Handling

WinRunner can react to exceptions (unexpected events and Defects) if the Developer defines an exception, writes an exception processing routine, and provides rules that will turn the exception handler on and off as required.

Test Results Analysis

WinRunner's Test Results File is based upon the Test Case.

Managing the Testing Process

WinRunner integrates with a separate Mercury Interactive product Program called TestDirector in order to manage the overall testing process.

Debugging

WinRunner has a visual debugger.

LoadRunner TestCenter

LoadRunner TestCenter is a global load-testing tool that enables organizations to manage multiple, concurrent load testing Projects across geographic locations. It Controls all aspects of large-scale load testing Projects, including resource allocation and scheduling, from a centralized location accessible via the Web.

Topaz

Topaz is an Application performance management tool. It is a solution that manages availability, performance and Service Levels from an end-user perspective. Topaz enables receipt of real-time alerts for early Warning of performance Issues and correlates end-user problems with Application data to accelerate problem resolution.

ActiveWatch

ActiveWatch is a service that manages availability, performance and Service Levels from an end-user perspective, from around the globe. ActiveWatch monitoring enables efficient addressing of IT Issues and provides users with performance opportunities.

ActiveTest

ActiveTest is a managed service that provides resources and experts to create Test Procedures that can load test Web-based Applications, validate performance and allows maximum scalability. The service reveals performance bottlenecks and capacity restraints within complex Applications before they impact End Users; and provides a summary Report with critical parameters such as number of concurrent users, response times, hits per second and throughput.

Astra LoadTest

Astra LoadTest easily tests the scalability and performance of Web Applications. It emulates the traffic of thousands of users to identify and isolate bottlenecks and optimize throughput.

Astra QuickTest

Astra QuickTest is an icon-based tool that allows Testers to validate dynamically changing Web Applications. It quickly creates interactive Maintainable tests by mirroring end-user behavior.

QuickTest Professional

QuickTest Professional is an icon-based tool that automates the Functional and Regression Testing of Web Applications. Mercury's ActiveScreen technology allows Testers to create reusable, easy-to-Maintain tests for browser-based, Java-based Applications and ERP/CRM solutions.

ProTune

ProTune is Production tuning software based on the technologies and processes employed by Mercury Interactive's Production tuning service, ActiveTune. Leveraging the methodologies and experience gained from thousands of ActiveTune engagements, ProTune offers the software solution for those Clients who wish to Develop their own Production tuning and optimization capabilities.

Product Evaluations

SDLC Support

These tools can be used in various segments of the SDLC, as shown in the following two charts:

	Development		Acceptance Testing					
	Unit	System	Functional	End-to-End	UAT	Load/Stress	Performance/Tuning	Business Integration Testing
LoadRunner	X	X		X		X		
XRunner	X	X	X					
ActiveTune		X		X	X	X	X (Production)	X
TestDirector	X	X						
Astra FastTrack	X	X						
WinRunner	X	X	X					X
LoadRunner TestCenter	X	X		X		X (global)		X
Topaz				X			Manages	
ActiveWatch				X			Manages	
ActiveTest		X		X		X	Validates	
Astra LoadTest	X	X		X		X		X
Astra QuickTest	X	X	X					
QuickTest Professional	X	X	X					X
ProTune		X		X	X	X	X(Production)	X

Figure 146 – Support During Development Projects

	Development		Acceptance Testing						
	Unit	System	Functional	Regression	End-to-End	UAT	Load/Stress	Performance/Tuning	Business Integration Testing
LoadRunner	X	X			X		X	X	X
XRunner		X	X						
ActiveTune		X			X	X	X	X (Production)	X
TestDirector	X	X							
Astra FastTrack	X	X							
WinRunner	X	X	X	X					X
LoadRunner Test Center	X	X			X		X (global)	X	X
Topaz					X			Manages	
ActiveWatch					X			Manages	
ActiveTest		X					X	X	X
Astra LoadTest		X					X	X	X
Astra QuickTest	X	X	X	X					X
QuickTest Professional	X	X	X	X					
ProTune		X			X	X	X	X (Production)	X

Figure 147 - Support During Maintenance Projects

Product Features

	Features					
	Enterprise Applications Tool	Web Applications Tool	Other Applications Support	Identifies Code Defects/ Run Time Defects	Identifies Untested Code	One Script for All Applications
LoadRunner	X	X	ERP/CRM	X		
XRunner	X			X		
ActiveTune	X	X	ERP/CRM			
TestDirector	X	X	ERP/CRM	X	X	
Astra FastTrack		X		X		
WinRunner	X	X	ERP/CRM	X		
LoadRunner TestCenter	X	X	ERP/CRM	X		
Topaz	X	X	ERP/CRM			
ActiveWatch	X	X	ERP/CRM			
ActiveTest		X	ERP/CRM	X		
Astra LoadTest		X	ERP/CRM	X		
Astra QuickTest		X	ERP/CRM	X		
QuickTest Professional	X	X	ERP/CRM	X		
ProTune	X	X	ERP/CRM			

Figure 148 – Product Features

Product Benefits

	Benefits							
	GUI Based Apps Test	Scripting Language	Script Dev Modules	Test Creation	Predictive Capability	Change	Verification	Single Point of Control
LoadRunner	X	X	X	X	X	X	X	X
XRunner		X			X	X	X	X
ActiveTune	Service	Service	Service	Service	X	X	X	X
TestDirector	X				X	X	X	X
Astra FastTrack	X					X	X	X
WinRunner	X	X	X	X	X	X	X	X
LoadRunner TestCenter	X	X	X	X	X	X	X	X
Topaz	X	X	X	X				X
ActiveWatch	X	X	X	X				X
ActiveTest	Service	Service	Service	Service	X	X	X	X
Astra LoadTest	X	X	X	X	X	X	X	X
Astra QuickTest	X	X	X	X	X	X	X	X
QuickTest Professional	X	X	X	X	X	X	X	X
ProTune	X	X	X	X	X	X	X	X

Figure 149 – Product Benefits

Product Functions

	Functions						
	Input	Validation	Tracking	Scheduler	Detail/ Summary Results	Process Management	Service Level Management
LoadRunner	X	X	X	X	X		X
XRunner	X	X	X		X		
ActiveTune	X	X	X	X	X		X
TestDirector		X	X	X	X	X	
Astra FastTrack			X		X		
WinRunner	X	X	X	X	X		
LoadRunner TestCenter	X	X	X	X	X		X
Topaz	X	X	X	X	X (with Real Time Alerts)		X
ActiveWatch	X	X	X	X	X		Availability
ActiveTest	X	X	X	X	X		X
Astra LoadTest	X	X	X	X	X		X
Astra QuickTest	X	X	X	X	X		
QuickTest Professional	X	X	X	X	X		
ProTune	X	X	X	X			X

Figure 150 – Product Functions

Empirex

e-Tester

e-Tester is an automated Functional and Regression Testing tool that uses point-and-click Visual Script™ technology that was introduced by Empirex in 1997. The Visual Scripts that are recorded by e-Tester are the common language used by all of the Components of e-Test Suite. The scripts can be reused in e-Load and e-Monitor without any translation.

The Visual Script process automates the process of creating Test Procedures and is complemented by ServerStats™ (for real-time Application performance data collection) and e-Reporter™ (for in-depth analysis of load and monitoring Test Results).

The primary product features are:

- Create scripts without any Programming knowledge
- Simplified creation of data driven tests
- Visual Test Results
- Test management and integration with Defect tracking
- Seamless integration between scalability, monitor, Test Management and Defect tracking
- Comprehensive solution

e-Load

e-Load is a robust Web load testing solution that allows testing the scalability and performance of Web (includes CRM/ERP) Applications. It can be used during

Application Development and post deployment to conduct Stress Testing. e-Load requires no Programming and can re-use Regression Tests without modification. A GUI with Visual Script™ technology reduces testing time and enables the creation of representative load scenarios.

The primary product features are:

- Test for both ramp and spike loads
- Improve load capacity throughout entire SDLC
- Pinpoint performance bottlenecks
- Both analysis and reporting tools

e-Manager

eManager Enterprise is a tool that allows users to organize, document, and manage the entire Web Application testing process. Its Interface and integrated management Modules (e.g., Requirements Module, Tests Creation Module, and Issues Module for Defect tracking) allow set up of a customized testing process and creation of a Detailed Test Plan that incorporates both manual and automated Test Cases.

In addition to the Modules, e-Manager Enterprise offers an integrated Reporting Interface to document/track all pertinent data points, and compile/present the data in various Reports for analysis to assess Application deployment readiness.

The primary product features are:

- E-Mail Notification of Defects
- Advanced grouping and filtering
- Web Interface
- Allows Clients to organize their Application testing efforts and bring a highly visible and repeatable framework to the web Application testing process

OneSight/FarSight

OneSight provides a way to manage performance of Web Applications and infrastructure. It tracks real-time performance of Web Applications, initiating and monitoring user transactions from multiple agent machines and monitoring the operation of Components such as servers and network devices to detect Changes that could affect Application performance. It exercises all levels of multi-tier Applications to detect Application performance bottlenecks.

FarSight is a hosted service that provides real-time Web site monitoring using objective performance data gathered from points on a global network. The service provides both URL and transaction monitoring from outside the firewall, from a secure network. It provides performance data/Reports for every object on a Web page, for every test and transaction and from every node.

Using OneScript™, a single Test Procedure that has been created to monitor Client transactions can be used without modifications with FarSight and throughout the Empirix testing/monitoring products and services.

The primary product features are:

- Assess website performance

- Flexibly modify web sites
- 24X7 monitoring
- Immediate alerts of any problems
- Interactive reports and graphs detailing tests
- Ability to pinpoint, recognize and isolate performance problems
- Focused on end-user experience

SDLC Support

	Development				Acceptance Testing				Implementation
	Unit	System	Functional	End-to-End	UAT	Load/Stress	Performance/Tuning		Business Integration Testing
e-TEST suite									
e-Tester		X	X	X					
e-Load		X		X		X	X		
e-Manager Enterprise			X						
OneSight									
FarSight									

Figure 151 - Support During Development Projects

	Development				Acceptance Testing					
	Unit	System	Functional	Regression	End-to-End	UAT	Performance/Tuning	Load/Stress	Business Integration Testing	Scalability Testing
e-TEST suite										
e-Tester		X	X	X	X					
e-Load		X			X		X	X		X
e-Manager Enterprise		X								
OneSight			X			X	X	Monitors		
FarSight			X			X	X	Monitors		

Figure 152 - Support During Maintenance Projects

Product Features

	Features					
	Enterprise Applications Tool	Web Applications Tool	Other Applications Support	Identifies Code Defects Run Time Defects	Identifies Untested Code	One Script for All Applications
e-TEST suite			CRM/ERP			
e-Tester		X	Wireless			
e-Load		X				
e-Manager Enterprise		X				
OneSight			X (and Web Infrastructure)			
FarSight			X (Global Network)			

Figure 153 – Product Features

Product Benefits

	Benefits							
	GUI Based Apps Test	Scripting Language	Script Development Modules	Test Creation	Predictive Capability	Change	Verif-ication	Single Point of Control
e-Tester	X	X (Visual GUI-based Procedures extendible with VBA)	X	X			x	x
e-Load	X	Uses same Procedures created in e-Tester			X		x	x
e-Manager Enterprise								
OneSight								
FarSight								

Figure 154 – Product Benefits

Product Functions

	Functions						
	Input	Validation	Tracking	Scheduler	Detail/Summary Results	Process Management	Service Level Management
e-TEST suite							
e-Tester	X	X	X	X	X		
e-Load	X	X	X	X	X		
e-Reporter			X		X		
e-Manager Enterprise			X	X	X	X	X
OneSight			X		X (with Real Time Alerts)		X
FarSight			X		X (with Real Time Alerts)		X

Figure 155 – Product Functions

McCabe

ReleaseRocketVerify

ReleaseRocketVerify is a tool that identifies, tracks, and Reports on the testing effort at the Change level. It focuses on the verification of Changes and testing results across the entire Application using real-time feedback.

IQ2 Suite

The IQ2 Suite tool manages changing software and helps the Development Manager ensure the reliability and stability of Applications. It merges the software Changes into released Code, and tracks Defect fixes/Changes.

SDLC Support

	Development			Acceptance Testing				
	Unit	System	Functional	End-to-End	UAT	Load/ Stress	Performance/ Tuning	Business Integration Testing
ReleaseRocketVerify	X		X					
IQ2 Suite		X						

Figure 156 - Support During Development Projects

	Development			Acceptance Testing						
	Unit	System	Functional	Regression	End-to-End	UAT	Performance / Tuning	Load/ Stress	Business Integration Testing	Scalability Testing
ReleaseRocketVerify	X		X							
IQ2 Suite		X								

Figure 157 - Support During Maintenance Projects

Product Features

	Features					
	Enterprise Applications Tool	Web Applications Tool	Other Apps Support	Identifies Code Defects/Run Time Defects	Identifies Untested Code	One Script for All Applications
ReleaseRocketVerify	X	X				
IQ2 Suite	X	X				

Figure 158 – Product Features

Product Benefits

	Benefits							
	GUI Based Applications Testing	Scripting Language	Script Development Modules	Test Creation	Predictive Capability	Change	Verification	Single Point of Control
ReleaseRocketVerify						X		
IQ2 Suite						X		

Figure 159 – Product Benefits

Product Functions

	Functions						
	Input	Validation	Tracking	Scheduler	Detail/ Summary	Process	Service Level
ReleaseRocketVerify	X	X	X		X		
IQ2 Suite				X		X	X

Figure 160 – Product Functions

Radview

WebLOAD

WebLOAD simulates real-life Internet user behavior and reports upon an Application's performance Issues to help companies successfully deploy high performing e-business Applications.

The primary product feature is the use of JavaScript for test agenda creation.

WebRM

WebRM is a Web Application testing and analysis solution that uses a GUI Interface to Plan and schedule tests manually or automatically as they become available and create test data results and Audit Reports. In addition, WebRM is Designed to work with MSDE (Microsoft Database E) and SQL Server.

WebFT

WebFT is a Web centric testing solution that efficiently verifies the Functional accuracy of Applications and tracks/Reports Defects to ensure that Applications perform as expected. It utilizes a GUI, automated dynamic script adaptation to Application Changes to support uninterrupted test runs, and an open Standards scripting language to reduce the amount of Developer/Tester Training that is required.

WebGTServices

WebGTServices provides services that help an organization test and deploy Web Applications quickly and efficiently. These services provide turnkey needs assessment and testing services that consist of:

- Development Test Planning
- Test Procedure Generation
- Test Execution
- Test Data Analysis

All of these products can be used on Windows, UNIX and Linux.

SDLC Support

	Development		Acceptance Testing					
	Unit	System	Functional	End-to-End	UAT	Load/Stress	Performance/Tuning	Business Integration Testing
WebLOAD		X					X	
WebRM							X	
WebFT			X					
WebGTServices		X	X					

Figure 161 - Support During Development Projects

	Development		Acceptance Testing						
	Unit	System	Functional	Regression	End-to-End	UAT	Performance/ Tuning	Load/ Stress	Business Integration Testing
WebLOAD		X					X		
WebRM									
WebFT			X						
WebGTS		X	X						

Figure 162 - Support During Maintenance Projects

Product Features

	Features					
	Enterprise Application Tool	Web Application Tool	Other Application Support	Identifies Code Defects/Run Time Defects	Identifies Untested Code	One Script for All Applications
WebLOAD		X				
WebRM		X	eBus/SQL			
WebFT		X				
WebGTServices		X				

Figure 163 – Product Features

Product Benefits

	Benefits							
	GUI Based Application Test	Scripting Language	Script Development Modules	Test Creation	Predictive Capability	Change	Verification	Single Point of Control
WebLOAD								
WebRM	X			X (shared)				
WebFT		X	X	X				
WebGTServices					X		X	

Figure 164 – Product Benefits

Product Functions

	Functions						
	Input	Validation	Tracking	Scheduler	Detail/Summary Results	Process Management	Service Level Management
WebLOAD			X	X	X		
WebRM				WebLOAD	X		
WebFT			X	X	X		
WebGTServices	X	X	X	X	X		

Figure 165 – Product Functions

Rational

Robot

Rational Robot is an object-oriented tool that allows users to create, modify, and run automated Functional and Regression Tests for e-Applications that are built on a wide variety of independent Development Environments.

The main product features are:

- Functional testing for Web, ERP, and Client/server Applications
- Regression Test
- Record and replay Test Procedures that recognize the objects in various Applications
- Track, report, and chart all the information about the quality assurance testing process
- Detect and repair problems associated with the various elements of a Web site
- View and edit the Test Procedures while recording
- Use the same Test Procedure, without modifications, to test an Application on multiple platforms
- Supports a range of Environments and languages, including HTML and DHMTL, Java, Microsoft Visual Basic and Visual C++, Oracle Developer/2000, Delphi, SAP, PeopleSoft, and Sybase PowerBuilder

TestManager

Rational TestManager is used to manage all of the aspects of testing and all of the sources of information that are related to the testing effort throughout all of the phases of a software Development Project, including:

- Test Planning
- Test Design
- Test Implementation
- Test Execution
- Result Analysis

The main product features are:

- Rational TestManager provides the ability to trace the impact of Requirements and other Changes by automatically flagging potentially impacted Test Cases. Rational TestManager is the central source of data and metrics on the quality-Status of the Project for the entire software team.
- With Rational TestManager, QA or QE managers, business analysts, Developers and Testers can each see Test Results from their own perspective, and use that information to make decisions pertinent to their work. Rational TestManager provides the entire team with continuous tracking of progress toward the goals of the Test Plan throughout the Development life cycle.

TeamTest

TeamTest automates Functional, regression, and highly scalable performance testing, and provides centralized Test Management and integrated Defect tracking.

The main product features are:

- ScriptAssure produces reusable Test Procedures and validates the interactive data that the Application generates
- Standard scripting language is Java
- Analyzes what business transactions, low-level Client calls, Application resources or Code Issues are causing bottlenecks. Virtual Testers for load testing are available separately.
- One centralized license server

Visual Test

Visual Test is an automated Functional testing tool that helps Developers and Testers rapidly create tests for Windows Applications that have been created with any Development tool. Rational Visual Test is integrated with Microsoft Developer Studio, a desktop Development Environment, and has extensive integration with Microsoft Visual C++.

The main product features are:

- Support for Microsoft J++ WFC Controls
- Winfo utility
- Suite Manager
- Batch-compilation all the Files in a Project into p-code
- Support for multiple monitors
- ActiveX/Web procedures
- RUNEX Function
- MSI-based installer

SDLC Support

	Development		Acceptance Testing					
	Unit	System	Functional	End-to-End	UAT	Load/Stress	Performance/Tuning	Business Integration Testing
Robot			X					
TestManager		X						
TeamTest		X	X			X		
ClearQuest		X						
PureCoverage		X	X					
Purify for Windows		X	X					
Purify for Unix		X	X					
Visual Test			X					
preVue		X					X	
Quantify		X					X	

Figure 166 - Support During Development Projects

	Development					Acceptance Testing				
	Unit	System	Functional	Regression	End-to-End	UAT	Load/Stress	Performance/Tuning	Business Integration Testing	Scalability Testing
Robot	X		X	X		X			X	
TestManager		X				X	X		X	
TeamTest		X	X	X		X	X		X	
ClearQuest		X								
PureCoverage		X	X							
Purify for Windows		X	X							
Purify for Unix		X	X							
Visual Test			X							
preVue		X		X				X		
Quantify		X						X		

Figure 167 - Support During Maintenance Projects

Product Features

	Features					
	Enterprise Application Tool	Web Application Tool	Other Application Support	Identifies Code Defects Run Time Defects	Identifies Untested Code	OneScript for All Applications
Rational Suite TestStudio			eBus/ERP			
Robot		X				X
TestManager	X	X	X			X
TeamTest	X	X	X			X
ClearQuest	X	X	X			
PureCoverage	X	X			X	
Purify for Windows	X	X		X		
Purify for Unix	X	X		X		
Visual Test	X	X				
preVue	X	X				
Quantify	X	X				

Figure 168 – Product Features

Product Benefits

	Benefits							
	GUI Based Applications Test	Scripting Language	Script Development Modules	Test Creation	Predictive Capability	Change	Verification	Single Point of Control
Robot	X	SQABasic or Java		X				
TestManager	X			X				X
TeamTest	X			X				
ClearQuest						X		
PureCoverage								
Purify for Windows								
Purify for Unix								
Visual Test				X				
preVue				X				
Quantify								

Figure 169 – Product Benefits

Product Functions

	Functions						
	Input	Validation	Tracking	Scheduler	Detail/ Summary Results	Process Management	Service Level Management
Robot	X	X	X		X		
TestManager			X			X	
TeamTest			X	X	X		
ClearQuest			X		X		
PureCoverage			X		X		
Purify for Windows			X		X		
Purify for Unix			X		X		
Visual Test	X	X	X		X		
preVue	X	X	X		X		
Quantify			X		X		

Figure 170 – Product Functions

Segue

SilkTest

SilkTest provides Functional Testing for e-business Applications, whether Web, Java, or traditional Client/Server based Applications. SilkTest also offers:

- Test Planning and management
- Direct Database access and validation
- Flexible Power Testing 4Test scripting language
- Built-in recovery Application for unattended testing
- Ability to test across multiple platforms, browsers, and technologies

In addition, SilkTest provides centralized testing of distributed Applications, distributed access to Test Results, extensive component testing, and cross platform Java testing.

The main product features are:
- Automated Function and Regression Testing
- Unattended Defect logging and recovery
- Object-oriented 4Test scripting language
- Cross-platform verification
- Component-based testing for heterogeneous Environments
- Centralized testing of distributed Applications
- Access Test Results from distributed Applications
- Manage software QA process from a central control point

Test Termination

SilkTest can terminate its process when an exception occurs which is not explicitly anticipated in the Test Case.

Add On Process and Extensions

The basic SilkTest can work with objects and windows that were built using the Standard Microsoft Foundation Class (MFC) library.

Object Hierarchy

SilkTest has a true object-oriented hierarchy of parent-child-grandchild-etc. relationships between windows and objects within windows.

Language Localization

SilkTest supports the single-byte IBM extended ASCII character set.

Database Verification

SilkTest has several built-in Functions that create SQL queries that can extract information from an ODBC compliant Database and save it to a File.

Data Driven Testing

SilkTest supports data-driven tests by building around an array of user-defined records. The test Code and the array must be hand Coded.

Recovery

SilkTest has a native recovery process that restores the Application to a stable condition (base state) when the test or Application fails.

Scripting Language

SilkTest has a proprietary, interpreted, scripting language. SilkTest also provides an effective object-oriented Programming language called 4Test.

Exception Handling

SilkTest can process exceptions (events and Defects) through its 4Test language.

Test Results Analysis

SilkTest's Test Results process is based upon the specific test that is run.

Managing the Testing Process

SilkTest has a built-in Function (SilkOrganizer) that establishes a Test Plan and can link the Test Plan to Test Cases. SilkOrganizer can also monitor the automation process and control the Execution of selected groups of Test Cases.

Debugging

SilkTest has a visual debugger.

SilkPerformer V

SilkPerformer V simulates the fluctuating load demands of a live e-business Environment, regardless of its size or complexity. The tool provides visual verification under varying loads for Web Applications. This allows for the capture of content Defects (including efficient root cause analysis/ identification) that occur in Production. In a single process, SilkPerformer V can predict the scalability and performance of Web Applications.

The main product features are:

- TrueLog, which is a process of visual verification under load
- TrueScale offers a process to replay transactions
- Intelligent recorder
- Visual customization for complex scenarios
- Support for a wide range of protocols and Interfaces used by typical Applications
- XML/DOM access
- Flexible workload models
- Built-in load-testing methodology
- Correlated infrastructure monitoring during a load-test

SilkVision

SilkVision is a web based enterprise-monitoring tool that helps users manage the performance and reliability of mission-critical e-business Applications. The product allows the user to define and schedule monitors that measure site availability and end-user experience on an ongoing basis. By collecting measurement data, the tool assists the Client in trending and capacity Planning, as well as in resolving performance and Functional Issues within the Production Environment. The visual logging capabilities support root-cause analysis.

The main product features are:

- Performer replay technology
- Flexible scheduler
- Sophisticated notifications (Pager, SMS)
- Integration with Enterprise Management Systems (EMS)
- Root-cause analysis based on TrueLog information from every location
- Route analysis through firewalls
- Results can be seen anywhere, through a Web-based front-end for reporting and Application Maintenance

- Monitoring of Application infrastructure through private agents

Platforms Supported:

	WINDOWS	UNIX
SilkTest	Yes	Yes
SilkPerformer V	Yes	No
SilkVision	Yes	No

Figure 171 – Platforms Supported by Segue

SDLC Support

	Development		Acceptance Testing					
	Unit	System	Functional	End-to-End	UAT	Load/ Stress	Performance/ Tuning	Business Integration Testing
SilkTest	X	X	X	X	X			X
SilkPerformer V	X	X	X	X	X	X	X	X

Figure 172 - Support During Development Projects

	Development		Acceptance Testing						
	Unit	System	Functional	Regression	End-to-End	UAT	Performance/ Tuning	Load/ Stress	Business Integration Testing
SilkTest	X	X	X	X	X	X			X
SilkPerformer V		X					X	X	X

Figure 173 - Support During Maintenance Projects

Product Features

	Features					
	Enterprise Application Tool	Web Application Tool	Other Application Support	Identifies Code Defects Run Time Defects	Identifies Untested Code	OneScript for All Application
SilkTest	X	X	X	X		X
SilkPerformer V	X	X	X	X		X

Figure 174 – Product Features

Product Benefits

	Benefits							
	GUI Based Application Test	Scripting Language	Script Development Modules	Test Creation	Predictive Capability	Change	Verification	Single Point of Control
SilkTest	X	X	X	X			X	X
SilkPerformer V		X	X	X	X		X	X

Figure 175 – Product Benefits

Product Functions

	Functions						
	Input	Validation	Tracking	Scheduler	Detail/ Summary Results	Process Management	Service Level Management
SilkTest	X	X	X		X		
SilkPerformer V	X	X	X		X		X
SilkPlan Pro				X	X	X	
SilkVision							X

Figure 176 – Product Functions

Summary

This chapter presented an evaluation of the tools that are available from several leading vendors along with an evaluation of their functionality, benefits, etc.

XVI. TEST OUTSOURCING

Outsourcing of processing was a major growth industry throughout most of the 1990's; however, in the last few years of the decade, the trend slowed as firms benefited from the expanding economy and concerns that complex outsourcing arrangements could not keep up with the pace of change. The economic slowdown over the last few years has encouraged firms to rethink this decision and to look at outsourcing for processing and for other areas such as testing as a necessary alternative. Some large firms now automatically evaluate support alternatives in three ways; Build, Buy or Outsource.

Forester Research predicted that 60% of large firms would use three or more service providers to support various aspects of their core processes after 2002. And, as IT managers re-evaluate the benefits of developing software in-house for non-differentiating Functions, they are increasingly concluding that acquiring software from vendors makes good business sense.

It is a natural extension for firms to look at third-party outsourcing as a potential solution for testing complex Applications. Some of the reasons for this are:

- Professional Testers can improve the quality of the Code and reduce the amount of time that it takes to test.
- Most businesses do not generate a continuous stream of large Projects that need sophisticated testing. However, the specialized skills that are needed would require these firms to keep people with these skills on their staff.
- Using a professional testing firm allows both firms to do what they do best. This also provides some independent review, which is not always the case when the Developers and Testers report to the same manager.
- By using a third-party specialist who knows the industry, firms can cut costs and reduce the amount of time that it takes to gear up for a major testing effort.
- If the testing is done off-site, this can also reduce the amount of space, hardware and other facilities that are required to support the peak load.
- If a vendor develops the software, the use of a third party tester can buffer the business from confrontational issues during testing.

Request for Proposal

Companies generally use a Request for Proposal (RFP) in order to evaluate alternative outsourcers. When Developing an RFP, managers should consider the points in the following table.

AREAS TO EXAMINE	SPECIFIC TOPICS
STAFF EXPERIENCE	Be sure to check on the experience of the people who will work on the Project, not just the experience that the company has had in the past. This includes: • Functions • Industries • Technologies Ask to meet / interview all of the key people who will be working on the Project.
STAFF GUARANTEES	Once you have selected certain people to work on your Project, you want to be sure that they will actually be assigned to you. Will the vendor guarantee that the people you want will be available for the Project (if they are still employed by the vendor)?
STAFF SKILLS	Verbal and written communication skills are critical in a testing Project. An interview can identify verbal skills, but in addition, you should review Reports that have been written by the Testers. Can they write clearly?

STAFF AVAILABILITY	Do you Plan to work overtime or on weekends? If so, will the people you need be available or will other commitments make this unlikely?	
STAFF BACKUP	Testing Projects inevitably have a tight timetable. Who will backup the people you select during: • Vacations • Illness • Termination	
STAFF TRAINING	Will their staff have to be Trained on your existing Applications or the new Application? Will they have to be Trained in any new testing or Project management software? Will they have to be Trained to operate in your technical Environment? Will they have to be Trained to use your SDLC or to work with your existing Quality Assurance process?	
IF THE EFFORT IS ON-SITE?	If the vendor will work at your location: • Will they have access to your hardware? How? • Is your hardware capable of supporting the various testing regions that will be required? • Will you have to obtain additional software licenses to allow the Testers to work on your equipment? • Do you have sufficient physical space that is conducive to effective work? • Are there sufficient PCs, telephones, faxes, copiers, printers, conference rooms, etc.? • How will you routinely communicate (Outlook, Lotus, etc.)? • What insurance do they have for their people working at your location?	
IF THE EFFORT IS OFF-SITE?	Will you have access to their hardware? Is their equipment the same as yours? • Printers • Modems • Networks • Etc. Will they be sharing the equipment that you need with another Client? Can their hardware support the multiple dedicated testing regions that you will need? What data communication facilities are available? Is the capacity sufficient? Is the line speed sufficient? Do they have a well-organized working environment? Do they have room for your people, if needed? Can they print from their Data Center to your printers? Do you need any special printers? Is their Data Center well organized, clean, efficient and secure? How does the Data Center handle back-ups? How long does it take to recover a back up? Are there any restrictions? Do they have the capacity to handle all of the volume you will need? Can they run multiple regions so that you can perform overlapping tests independently? What is the availability of their hardware on weekends? Do they have the same hardware that you have? What are the working conditions for the testing staff? Over crowding leads to Defects and lower productivity. What insurance do they have to protect your activities at their location? How do they handle security: • Physical • Network • Personnel	
TESTING SOFTWARE	What testing software will be used? Are there a sufficient number of user licenses? Will this account for simultaneous users? Who will pay for it? Will the Testers also need software for spreadsheets and a Database (e.g., Excel and Access)? If so, how many licenses are needed and who will pay for them?	
PROJECT MANAGEMENT	What Project Management software will be used? Is the Issues Tracking Process well defined? Who will pay for the licenses? What SDLC will you use? When working with a third party software provider or testing firm, any mismatch in expectations could lead to significant problems.	
ORGANIZATION	How will the vendor's team be organized, and to whom will they report? Will different people from the vendor's team report to different people on your staff?	

	Will the vendor's team need a dedicated person to do management Reporting?
	Will your Test Team need a dedicated person to coordinate all of the technology support that is required?
REPORTING	Agree in advance on what is included in the Reports, including: • Content • Format • Timing
	Identify in advance what Documentation will be needed for Audit, including: • Correspondence • Contracts • Issues Tracking Reports • Decisions
	Identify what will be covered in the Test Plan and who will Develop each portion of the Test Plan, including: • Format • Review/Approvals • Process Identify who will Develop the Test Cases.
	See samples of the testing firm's output, including: • Test Plan • Test Conditions • Test Cases with Expected Results • Issues Tracking and related correspondence • Project Updates • Project Summary Recommendations

	Define in advance how you will communicate with an off-site vendor, including:
	• Face to Face
	• Written Reports
	• Teleconferencing
	   – Group
	   – Individual
	• Shared electronic White Board

	For Issues Tracking, establish a single shared data base that reports: • Problems (including a related File of examples) • Resolutions • Status • Re-Test Results
	Define the Project's end deliverables. As the Project comes to an end there should be clearly defined Project end deliverables. The deliverables may include the following: • Final updated Project Plan • Completed Audit material • Summary letter of outstanding issues and recommendations from the vendor
COST	Personnel costs will be the largest single cost in a Test Plan, so the Test Manager must be extremely careful in defining the costs, including: • Flat rates • Daily / Hourly rates • Weekend/Overtime/Holiday rates • Severance agreements • T&E reimbursement
	If hardware is going to be provided by the vendor, specifically identify the parameters of cost, including: • Flat rates • Daily / Hourly rates • Weekend/Overtime/Holiday rates • Costs for additional, unplanned hardware
OTHER TERMS AND CONDITIONS	In the contract, there are several specific terms that should also be addressed, including: • Payment terms • Performance bonus • Termination • Extra charges • Retainer for future Regression Testing • Fees for late deliveries from the vendor • Fees for late payments to the vendor

Figure 177 – Outsourcing RFP Points

Outsourcing Guidelines

When selecting an outsourcing firm to conduct testing, a Test Manager must follow a highly structured process to ensure success. Some of the areas to consider are shown in the following chart.

Category/Topic	Do	Don't
Relationships	When possible, start with a low Risk Function.	Do not wait until you are well into the Test Process before you see any results.
	Use face-to-face meetings early in the process when people need to get to know each other.	
	Educate the vendor in your: • Company • Products • Customers • Application	Do not assume that the vendor's staff knows the Application domain just because the vendor's salesmen say so.
Process - SDLC	Have a detailed SDLC that has been agreed to in advance and which is followed.	Do not skip steps
	Ensure that Users and IT staff have been Trained in the use of the same SDLC. Different SDLCs have different procedures, and misunderstandings could occur.	
Requirements	Have clear, detailed Requirements that precisely define what is needed.	Do not assume that the Developers will know what is needed.
	Ensure that Testers understand the Requirements before they begin Developing Test Cases.	
	If testing is conducted remotely, have someone at each location responsible for communication.	Do not rely solely on written comments.
Management Contacts	Establish clear contact points, and define who can talk to whom (Consider having a single point of contact).	Do not take shortcuts.
	Identify in advance, in writing, what must be delivered before a Task can proceed.	
	Document all meetings.	
	Document all verbal agreements, approvals, and decisions.	
	Establish contact lists with telephone numbers and email addresses.	
Decisions	Define who in each organization has the authority to make each level of decision.	Do not assume that a decision maker will always be available when they are needed – establish a backup.
	Define in advance the required timeframe for decisions by each organization.	
Meetings	Define who will be required to attend regular update meetings.	Do not schedule wall-to-wall meetings. Do not hold meetings just because they are routinely scheduled.
	Define who will be in charge of each Meeting.	
	Limit participants to those who can contribute.	
	Establish a clear agenda for each meeting	Do not deviate from the agenda
	Consider regular Video conferencing for meetings with remote staff.	
Staffing	Each team must have a qualified Business Analyst who knows the industry.	Do not assume that people know everything about their areas of responsibility.

	Ensure that there are a sufficient number of people who know the various Applications, with backup for key people.	Do not assume that people know multiple Applications or processes, test them.
APPLICATION KNOWLEDGE	Allocate time early in the Test Project Plan to learn about the Applications: • Business Purpose • Interfaces • Critical Timing • Critical Data • Impact of Defects • Logical Flow (Modules)	
	Ensure that there is sufficient Budget to pay for the required Training.	
TECHNICAL KNOWLEDGE	Define the skills required in advance and assign people with the required skills.	Do not assign people with inadequate/inappropriate technical skills with the hope that they will learn.
ORGANIZATIONAL KNOWLEDGE	The Outsourcing staff should know and understand the Customer's organization.	
BUSINESS KNOWLEDGE	Ensure that each of the assigned staff has a fundamental understanding of the business process.	Do not require that everyone on the testing team have the same level of knowledge or skill.
CROSS TRAINING	Every person should have a backup.	Do not assume Application knowledge, test for it.
	Allocate time / Budget for cross Training.	
REPORTING	Define the Scope and details of Reporting in advance.	
	Ensure that Reporting occurs regularly to the IT management as well as Business Management.	Do not assume that senior management knows about problems.
	Create 'strawman' Reports to get concurrence in advance to the content and format.	
	Allocate time in the Test Project Plan for each appropriate person to assemble the required information.	Do not assume that people can prepare Reports while they are testing. Allocate specific time for this Function.
	Assign a person (part-time or full time) with the primary responsibility to manage Reporting, and assign and Train a backup.	Do not assume that the Project Manager will be able to assemble the regular Reporting as a part of their job.
	Ensure that there is sufficient time/Budget to pay for the required Reporting.	
COST	Define the activities that are included in any Fixed Cost portion of the contract.	
	Define the hourly/daily rates that will be applied for activities that go beyond the Fixed Cost contract	
PAYMENTS	Define how invoices will be prepared, when they will be submitted, to whom, and how fast they will be paid.	
	Define a late fee.	
QUALITY	Define the quality Standards in advance.	
	Define the quality goals in advance.	
TIMELINESS	Define the timeliness required for various categories of Development / Maintenance / Defect resolution, etc.	

Figure 178 – Test Outsourcing Do's and Don'ts

Summary

This chapter discusses the opportunity to outsource Testing Projects, and provides one checklist to evaluate potential outsourcers and another checklist to manage the outsourcing process.

Appendix I - Test Team Job Descriptions

Staffing the Test Team is critical to the success of any testing effort. In a Large Project there may be multiple people assigned to each position, while in a Medium or Small Project, the job descriptions of several of the positions may be combined and assigned to one person. One exception to this is the role of the Test Team Manager since there should be only one person who is ultimately responsible for the overall testing effort.

However, no matter what the Project is, there are two key areas of experience that will need to be covered:

- Testing Experience
- Business Experience

Very often, the candidates who are being recruited for the testing Project will not have the ideal expertise for their position in both of these areas. It will be important to the success of the Project to ensure that the Test Team as a whole has the depth of experience in both these areas, even if each team member does not have both.

The structure of the Test Team will vary based on the size and type of testing effort. The following is one example of the structure:

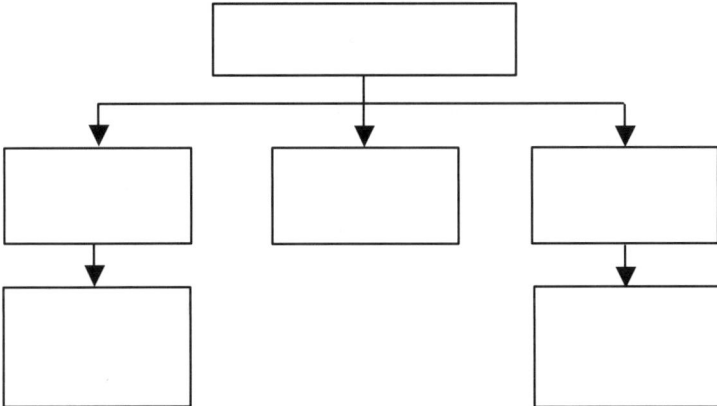

Figure 179 – Sample Organizational Structure

The structure of the Test Team will vary by organization, and needs to be customized for each Project. It is possible that the Project will require multiple Test Section Managers and it may make sense to have the Test Tool Administrator report to the Test Team Leaders for each group.

The following are suggested job descriptions for the members of the Test Team who were identified in Chapter II – Key Players in the Testing Process.

Test Team Manager

Responsibilities

- Participates in the Implementation of the Application from Planning through post-Production support
- Responsible for the overall quality of the Application and Documentation
- Responsible for establishing and monitoring the Test Team Budget
- Hires, Develops and provides Training for Test Team
- Establishes the software quality assurance goals
- Provides Guidelines to support the testing and quality practices
- Provides the testing process, methods and Standards to the Test Team
- Establishes the test goals
- Determines the testing strategy and approach for all Application Components
- Develops or approves the Test Plans and Test Procedures
- Oversees the Test Execution and results sign-offs
- Performs periodic reviews to ensure that the company-defined Test Processes, methods and Standards are being followed
- Communicates the Defects and Issues discovered during testing and assists in their resolution as needed
- Provides recommendations on how to proceed based upon the Test Results
- Provides Status reporting to all interest parties

Skills/Experience

- High degree of expertise in analysis, testing, data access/retrieval and problem solving
- Must be highly self-motivated
- 7-10 years testing experience
- 3-5 years experience in Test Team management
- Degree in Information Technologies or related degree
- Knowledge of the business supported by the Application
- Knowledge of the technology used by the Application
- Positive attitude
- Creative thinker
- Effective skills in:
    - Communication
    - Leadership/motivation
    - Problem solving

Test Section Manager

Responsibilities

- Assists in the Development of the Test Plans and Test Procedures
- Performs reviews of the Test Teams work to ensure company-defined processes, methods and Standards are being followed
- Communicates the Defects and Issues discovered during testing and assists in their resolution as needed
- Assists in problem resolution
- Provides Status reporting to the Test Manager and Test Team
- Maintains repository of all test Documentation related to their activities
- Provides Training for Test Team
- Co-ordinates with other Section Managers to maximize the testing efforts

Skills/Experience

- High degree of expertise in analysis, testing, data access/retrieval and problem solving
- Requires minimal supervision
- 5-7 years testing experience
- 1-2 years experience in Test Team management
- Degree in Information Technologies or related degree
- Knowledge of the business supported by the Application
- Experience in the technology supporting the product
- Positive attitude
- Creative thinker
- Effective skills in:
  - Communication
  - Leadership/motivation
  - Problem solving

Test Administrator

Responsibilities

- Maintains repository of all test Documentation including: Test Plans, Test Procedures, Expected Results and Actual Results
- Maintains the Issues Database to track and report on all Issues
- Works with interested parties to assign priority and resolution of all Issues
- Maintains and report on the Project Plan
- Creates/Maintains the Test Library for future testing efforts

Skills/Experience

- High degree of expertise in analysis, testing, data access/retrieval and problem solving
- Requires minimal supervision
- Degree in Information Technologies or related degree
- 2-4 years experience in Project Management and/or testing
- Knowledge of the business the Application supports
- Experience in the technology supporting the product
- Positive attitude
- Creative thinker
- Effective Communication skills

Testers

Responsibilities

- Designs, writes, documents and Maintains Test Procedures and Expected Results
- Executes, evaluates and documents Test Results
- Documents Issues during all phases of testing
- Works with Technology to replicate, resolve and Re-test Issues
- Reviews Application Training or User materials for accuracy

Skills/Experience

- High degree of expertise in analysis, testing, data access/retrieval and problem solving
- Requires minimal supervision
- Degree in Information Technologies or related degree
- 2-4 years experience in testing
- Knowledge of the business supported by the Application is preferred
- Experience in the technology supporting the product
- Positive attitude
- Creative thinker
- Effective Communication skills

Test Tool Implementer

Responsibilities

- Participates in evaluating and selecting automated tools for testing
- Executes and documents results of Test Procedures using the tool
- Documents Issues during all phases of testing
- Works with Technology to replicate, resolve and Re-test Issues
- Trains Testers on the use of the tool

Skills/Experience
- Requires minimal supervision
- 2-4 years experience in testing
- Degree in Information Technologies or related degree
- Knowledge of the business supported by the Application a plus
- Experience in the Test Tool technology
- Positive attitude
- Creative thinker
- Effective Communication skills

Appendix II - Automation Vendors

Auto Tester

Corporate Office

Street Address	10300 N. Central Expressway / Suite 550
City	Dallas
State	TX
Zip	75231
Phone	800-328-1196
Web Site	www.autotester.com
Product List	AutoTester ONE
	AutoTester ONE Special Edition for SAP
	AutoController

Primary Marketing Office

Dwight Kallstrom

10300 N. Central Expressway, Suite 550

Dallas, TX 75231

214-368-1196, ext 205

dwightk@autotester.com

Compuware

Corporate Office

Street Address	31440 Northwestern Hwy
City	Farmington Hills
State	MI
Zip	48334
Phone	800-521-9353
	248-737-7300
Web Site	http://www.compuware.com/

Product List

- QACenter
  - QARun
  - QALoad
  - TestPartner
  - QADirector
    - Reconcile
    - TrackRecord
  - PointForward
- DevPartner
  - DevPartner Studio, DevPartnerDB, Driver Studio, SoftICE
- Strobe
  - CICS, DB2, MQ
- Vantage
  - Client Vantage, Server Vantage, Network Vantage, Application Vantage, Application Expert
- Xpediter
  - CICS, IMS, TSO

Primary Marketing Office

Doug Anter, John Lilienthal

31440 Northwester Highway

Farmington Hills, MI 48334

800-521-9353

john.lilienthal@compuware.com

doug.anter@compuware.com

Mercury Interactive

Corporate Office

Street Address	1325 Borregas Ave.
City	Sunnyvale
State	CA
Zip	94089
Phone	800-TEST911
Web Site	www.merc-int.com

Product List LoadRunner, Xrunner, ActiveTune, TestDirector, Astra FastTrack, WinRunner, LoadRunner TestCenter, Topaz, ActiveWatch, ActiveTest, Astra LoadTest, Astra QuickTest, QuickTest Professional, ProTune

Primary Marketing Office LoadRunner, WinRunner, TestDirector, ActiveTest, Astra LoadTest, Astra QuickTest, Astra FastTrack , ERP/CRM

Name	Judith Meinhalt
Street Address	1325 Borregas Ave.
City	Sunnyvale, CA 94089
Telephone	408 822-5557
Email	jmeinhalt@merc-int.com

Primary Marketing Office Topaz and ActiveWatch

Michele Holguin

1325 Borregas Ave.

Sunnyvale, CA 94089

408 822-5389

mholguin@merc-int.com

Primary Marketing Office ActiveTune, ProTune

Danielle Hamel

1325 Borregas Ave.

Sunnyvale, CA 94089

408 822-5389

dhamel@merc-int.com

Empirex

Corporate Office

Street Address	1430 Main Street
City	Waltham
State	MA
Zip	02451
Phone	781-993-8500
Web Site	www.empirix.com

Product List

- e-Tester
- e-Load
- e-Reporter
- e-Manager Enterprise
- OneSight
- FarSight
- e-TEST Suite

Primary Marketing Office

Karen Drake

1430 Main Street

Waltham, MA 02451

781 993-8619

kdrake@empirix.com

McCabe

Corporate Office

Street Address	5501 Twin Knolls Rd.
City	Columbia
State	MD
Zip	21045
Phone	800-638-6316
Web Site	www.mccabe.com
Product List	IQ2 Suite
	ReleaseRocketVerify

Primary Marketing Office

Julie McIntyre

5501 Twin Knolls Rd.

Columbia, MD 21045

800 638-6316 x 204

jules@mccabe.com

Radview

Corporate Office

Street Address	7 NE Executive Park
City	Burlington
State	MA
Zip	01803
Phone	781 238 1111
Web Site	www.radview.com
Product List	WebLOAD
	WebRM
	WebFT
	WebGTS

Primary Marketing Office

Laura Naylor

7 New England Executive Park

Burlington, MA 01803

781 238-1117

lnaylor@radview.com

Rational

Corporate Office

Street Address	2800 San Tomas Expressway
City	Santa Clara
State	CA
Zip	95051
Phone	408-496-3600
Web Site	www.rational.com

Product List

- Robot
- PureCoverageWindows
- PureCoverageUnix
- Purify for Windows
- Purify for Unix
- TestManager
- TeamTest
- ClearQuest
- Visual Test
- preVue
- Quantify

Primary Marketing Office

Name	Dave Locke
Street Address	2800 San Tomas Expressway
City	Santa Clara, CA 95051
Telephone	303 544-7375
email	dlocke@rational.com

Segue Software

Corporate Office

Street Address	201 Spring Street
City	Lexington
State	MA
Zip	2421
Phone	800-287-1329
	781-402-1000
Web Site	www.segue.com

Product List

SilkTest
SilkPerformer V
SilkScheduler
SilkPlan
Segue Software

Primary Marketing Office

Name	Genevieve Cross
Street Address	201 Spring Street
City	Lexington, MA 02421
Telephone	781-402-5880
email	gcross@segue.com

Appendix III - Definitions Used to Compare Testing Tools

In Chapter XV, several different tools from different vendors were analyzed and compared. The following definitions were used in the preparation of the Tables.

Feature Attribute Definitions

Other Applications Support

The tools can list other, specifically supported, Applications in addition to either Enterprise or Web Applications supported by the testing tool.

One Script for All Applications

The testing tool's Test Scripts can be utilized to test either all Enterprise or all Web Applications.

Benefit Attribute Definitions

GUI Based Applications Testing

In this case, the testing tool offers a Graphical User Interface for the Tester; providing the use of buttons, icons, and graphical information instead of only text, which facilitates ease of use of the testing tool.

Scripting Language

The testing tool provides Developers/Testers with the capability to quickly and productively create and execute repeatable Test Procedures, verify tests and analyze Test Results.

Script Development Modules

The testing tool provides Testers with Modules that facilitate establishing Application Test Procedures. These Modules allow Testers to record the traffic an Application generates and convert it into a Test Procedure.

The resulting scripts directly reflect the actual traffic generated by the Application and measure the time taken to perform required transactions to ensure that the Application under test will perform to specification under load.

Test Creation

The testing tool provides Developers/Testers with an automated, as opposed to manual, capability to create various required Test Procedures through the use of the tool's scripting language Modules for rapid setup and Execution of the scripts.

Predictive Capability

The testing tool's monitoring of the entire Application from the End User through the network (server and Database tiers), tracks status at each point and logs results; thereby enabling the capability to predict Application performance.

Appendix IV - Software Acceptance Criteria

The following is a sample acceptance criterion for the case of a vendor supplied Application or Enhancement. The details must be negotiated by the buyer and the seller, and will vary based upon the specific Conditions of the Project. These Conditions can be included in the base contract or in a separate Service Level Agreement.

When another firm is doing the Development, a testing team should have a Document that defines the relationship between the two firms. There are three areas that can be defined:

Initial Software Delivery

Software Developed by a vendor should be delivered according to the agreed upon schedule. When a component of Code is delivered late, it could be subject to a penalty of 5% of the cost for that component, per day.

A component is defined as a deliverable that relates to a single specification, and would typically consist of an Interface or the delivery of a Function.

Code Repair Timeliness

The buyer will identify software Issues during testing, and with input from the vendor, the buyer usually will prioritize them in one of the following categories:

CATEGORY	DEFINITION*	RESPONSE PERIOD**	REPAIR PERIOD***	PENALTY FOR 1-5 DAYS LATE	PENALTY FOR 6+ DAYS LATE
Critical	A Critical Issue is one that prevents the individual test from continuing and/or prevents a large number of downstream Components or scripts from being Executed. A critical Defect is one that will impact the ability to Maintain the Test Calendar.	24 hours	24 hours	$5K/day	$10K/day
High	A High Issue is one that must be resolved in order to begin live Production but does not prevent the testing from continuing.	24 hours	2 Business Days	$3K/day	$10K/day
Medium	A Medium Issue is one that must be fixed, but for which there is a manual workaround that can be used to begin Production	72 hours	Next release	$1K/day	$5K/day
Low	A Low Issue is one that the business would like to have fixed.	10 Business Days	Next release	None	None

Figure 180 - Sample Software Acceptance Criteria

* An Issue is defined as a single Defect in the Code that is delivered, and will normally map to a single Issue that has been reported in the Issues Tracking System.

** The Response Period is the period between the time the vendor is notified of the Defect and the total time it will take the vendor to confirm that the Defect does exist and to identify how the Issue will be resolved.

*** The Repair Period is the time that the vendor has to resolve and load the repaired Code onto the test Application so that it is available for Re-testing by the buyer. This is measured from the time the Issue has been accepted by the vendor and the time the vendor states the code is ready for re-testing.

Code Quality

After an Issue has been identified to the vendor, each repair is expected to be correct, complete and to not cause another Defect. If corrected Code does not pass a Re-test, the following quality penalty will be in effect:

ERRONEOUS REPAIR	PENALTY PER ERRONEOUS REPAIR
Discovery of initial Defect	--
First delivery of repaired Code	$ 10K
Second delivery of the same repaired Code	$ 20K
Each subsequent erroneous delivery	$ 50K

Figure 181 - Possible Penalties

Service Level Agreement

During all phases of the buyer's Acceptance Tests, the vendor must be able to demonstrate that each Change has been thoroughly Unit Tested and Systems Integration Tested by having Documented Expected and Actual Results. The buyer should be given immediate access to these Documents, if requested.

Since Applications are typically delivered in phases, this Acceptance Criteria covers Defects that are found in any Code (original or Changes) during any phase of the Project. This means that if after the first phase goes live, the buyer finds a Defect in the Code that has already been delivered and which is already in live Production, the vendor will continue to make the repairs according to the criteria in this Document.

The vendor agrees to provide access to appropriate vendor staff during testing to define and resolve Defects.

Appendix V - Case Studies

Case Study – Compuware / AZUR-GMF [54]

Industry

Insurance

Challenge

A leader in web-based insurance sales, French based AZUR-GMF sought to rationalize its Internet architecture by choosing between two servers with differing operating systems. It needed automated testing tools to help make this decision and roll out a new system.

Solution

The integration between Compuware's automated testing tools (QA Center[55]) and its server and network performance management tools (Vantage) lets AZUR GMF create realistic load tests, identify bottlenecks and reach production objectives on time.

Tools Used

- QACenter Performance Edition
- ServerVantage
- NetworkVantage

Background

After setting up its e-commerce division, the company decided to offer all its products online, including liability insurance, life insurance, investments, retirement and legal protection, as well as creating a space for exchange and for provision of Client services, advice and other information. This put a premium on a high level of performance for its web infrastructure. Since 1999, AZUR-GMF has employed an Internet architecture based on two servers, one running in Microsoft NT and the other in WebSphere NT/MTS. In an effort to harmonize and rationalize this architecture, the company opted for a single operating system: WebSphere.

The IT systems manager decided to test this decision proactively, before deployment, to ensure that the system the company's business depended on would be robust and optimized to receive, manage and process inquiries in a rapid and transparent manner. The choice was narrowed down to two vendors: Compuware and a major competitor. After carefully assessing the two offers, AZUR-GMF chose Compuware because of its close adherence to the specifications, its rapid response time and, above all, the integration of Compuware's testing tools with its Vantage suite of performance management tools.

"Compuware performed better in terms of response time and performance follow-up, and appeared to us to be more suitably placed to help us benefit from its experience in incremental load tests and analysis, thanks to the integration of its different offerings," commented Dominique Face, Director of Architecture, Support and Development Tools at AZUR-GMF. "Also, its offer of professional services resources guaranteed access to proven experts if difficulties arose in setting up the Project." Once the choice was made, representatives of AZUR-GMF's IT department went to work. "Two days were enough for full training in the different Compuware solutions," Face says. "The comments made by our technicians after the training sessions were encouraging. They found the learning and Implementation phases very straightforward."

---

[54] Reprinted with permission from Compuware

[55] QA Center is a collection of Compuware tools, including QA Run, QA Director and FileAID/CS.

Finding Bottlenecks

Pre-production is the last chance to verify and validate technical choices made in the deployment phase, before users get access to the system. Thanks to in-depth automated testing, AZUR-GMF can simulate the performance of a particular system without involving real users or their equipment. Load tests are performed by creating virtual users who make various predetermined requests as many times as necessary. Increasing the number of virtual users helps establish the point at which system performance slows down or is interrupted. Using Compuware's QACenter Performance Edition, AZUR-GMF could predict system performance under different load conditions and thus determine realistic expectations. By combining QACenter and the Vantage tools, AZUR-GMF could:

- Simulate multiple users realistically and thus gain a precise idea of response times (QACenter)
- Collect information concerning system availability and how performance was affected by load (ServerVantage)
- Learn how different systems consumed network resources (NetworkVantage).

These tests began by simulating 100 simultaneous users, escalating to 250 and then 500. The tests quickly identified a bottleneck in the SNA communications section at only 12 simultaneous requests - far below the target of 100. The tests generated a detailed analysis of the problem that identified the factors responsible for the slowdown. A simple adjustment of SNA access to the mainframes eradicated the bottleneck and made 100 parallel requests possible.

The tests allowed AZUR-GMF to identify http setting problems that, once corrected, increased simultaneous access from 60 to over 100 users. Testing was also extended to the company's intranet, which is more and more crucial to employee productivity.

> "Compuware performed better in terms of response time and performance follow-up, and appeared to us to be more suitably placed to help us benefit from its experience in incremental load tests and analysis, thanks to the integration of its different offerings."
> **Dominique Face, Director of Architecture, Support and Development Tools. AZUR-GMF**

Postponing Costly Upgrades

The tests with QACenter are helping AZUR-GMF reach production objectives. They could also help save money, because load testing enables the company to validate when hardware needs to be replaced. Currently it uses two Compaq DL 580 servers, dual-processor servers with one GB of RAM, to host all its web sites. Once the new sites are fully in production, simulations of expected activity will determine whether the company needs to upgrade to two quadriprocessor servers with 4GB of RAM, a sizeable investment. With automated testing and performance management using Compuware tools, the IT managers at AZUR-GMF will have hard evidence of whether an upgrade is necessary and can avoid any unneeded expense. You might say that Compuware tools are their insurance policy.

Case Study – Compuware / CIBC [56]

Industry
Banking

Challenge
To reduce the amount of time and money spent on testing throughout the organization

Solution
Acquire and utilize an integrated testing tool set from a single vendor.

---

[56] Reprinted with permission from Compuware

Tools Used

QARun

Background

Every consumer can point to a time when they saved money by buying in bulk and obtain some economies of scale. The more you buy, the more you save. The same holds true for large corporations. Take the Canadian Imperial Bank of Commerce (CIBC), for example. With 47,000 employees serving 6.5 million Clients around the world, CIBC is one of North America's leading financial institutions.

From personal banking to corporate and investment banking, CIBC provides a wide range of financial services to Clients worldwide. Being successful, however, has its costs. That's why CIBC is developing eSource, a web-based procurement system for operating resources. And that's also why they turned to Compuware and its QARun automated testing product: to save significantly on software testing costs, and gain another economy of scale. CIBC found that the more they use QARun to test, the more time - and money - they save.

The Search for Savings

Because financial institutions such as CIBC do not purchase any direct materials, all of the goods and services they acquire - from pencils to PCs, travel expenses to contracts—are operating resources. By 1998, the cost to CIBC of purchasing operating resources had reached more than US$880 million. To reduce costs, CIBC began in 1998 to develop eSource, a bank-wide enterprise resource planning (ERP) initiative for automated procurement. The eSource system is expected to save US$40 to $60 million annually once it is rolled out to all CIBC regions in Canada and the US. There was another, very important economy of scale to be found before eSource could be brought online - how to assure the system's quality, productivity and Y2K readiness within a relatively short amount of time? Comprehensive Test Procedures for the system had to be written and executed to determine how well eSource would work, or where potential production problems may lay. The time set for testing in pre-production was very tight. Bringing a poorly performing system on line was simply not an option.

> "QARun ensures consistency of input and ensures the output is based only on data from the application. For large-scale Application testing, I can't see a tool that could be applied more easily than QARun"
> **Ray Grech Director of Application Services Canadian Imperial Bank of Commerce**

The answer: Compuware QARun, a powerful tool for automating test creation, execution and analysis tasks for any system that can be accessed through the Microsoft graphical User Interface. A key QARun feature for testing eSource is the central repository that stores all Test Procedures, checks, events and object definitions. The repository made it easy for the testing staff to modify and maintain test assets—important for test re-usability. By using QARun, CIBC saved considerable amounts of time and labor to bring eSource online successfully. "QARun is a very powerful testing tool," said Shelley Tam, a Director of Global Technology in the CIBC Global Technology division. "The more you use it, the more you like what it is able to accomplish."

A Solution Is at Hand

Begun in August 1998, the eSource Project melds together two software packages. A web-based User Interface from Ariba handles Purchase Requests, Approval Processing and

> "Due to the true repository based design and architecture of QARun, which maps the scripts, checks objects, events and logs in a database, the scripts are rendered more re-usable than the competition. In a matter of hours, the scripts can be altered and/or re-sourced to work with a new version of the Application being tested. This saves time and the effort of re-scripting the old scripts to match the new version of the Application. QARun is Maintainable!"
> **Mike Amen, Cincinnati, OH**

Receipts, while a PeopleSoft program handles Purchase Orders, Accounts Payable and Asset Management, for back-office processing. "Once completed, eSource will have about 1,000 users in Canada, and 300 to 400 in the United States," said Bill Kim, a Director of Global Technology who led the User Acceptance Testing (UAT) and the QARun for Y2K Test Team at the eSource Project. "With a Project this complex, we really needed a good automated testing tool. However, because we had already used QARun for Y2K testing at CIBC World Markets, we were already familiar with its capabilities, and the Project management suggested that we use it to test eSource as well."

"Before we started our Y2K work a couple of years ago, senior management asked us to evaluate testing tools that would synthesize, read and manage the entire test base," said Tam. "We needed a tool that would replace the repetitive and time-consuming tasks related to testing. We first chose Compuware File-AID to perform File Conversions and data comparisons, and we soon found out that QARun would work very well with File-AID to help us generate test data and test scenarios, execute comprehensive tests and identify defects. It has become our best testing tool, and a standard within CIBC World Markets."

Fast, Accurate Test Results

Before testing on eSource began, CIBC brought in one QARun consultant from Compuware. In two weeks, the team was sufficiently trained on QARun, and began their Y2K testing. "We had about 75 scripts that represent approximately 1,000 screens' navigation, to perform Y2K testing on eSource," said Kim. "And the development of all Test Procedures took only one week by the three freshly trained team members. The most fun part of the testing was watching the QARun test execution, which only took one and one-half hours to complete. Another significant time saver was using QARun to build test data. Without QARun, we would have needed many key operators entering in data for days."

"The ability to take advantage of the easy-to-use QARun tool paid big dividends during eSource Y2K testing," said Ethel Clarke, the Overall Testing Manager of eSource. "In fact, it would have been very difficult if not impossible to maintain schedule dates without it."

Accuracy of test results is a strong feature of QARun - and one CIBC relies on. "QARun eliminates the concern of human defect when we're doing testing," noted Tam. "We trust the accuracy of QARun tests." Kim described another example of the savings that QARun afforded CIBC.

"Because the Y2K testing using QARun is so quick, we were able to share the test environment, the UNIX machine, with UAT. That is, to apply current system dates for UAT during the day and apply Y2K dates as system dates during the evening for Y2K tests. Without QARun, we would have had to buy another UNIX server. That's a big savings right there."

Testing in a Fraction of the Time

With QARun now working as a testing standard at CIBC World Markets, its benefits in saving time and labor are becoming readily apparent in other Projects as well. Especially to Ray Grech, director of system services, who said, "My first experience using QARun was to test a large number of Excel spreadsheets used by the back office," said Grech. "There were 1,200 spreadsheets for the CIBC Core Operations group that needed to be tested. We asked a large consulting firm to estimate how long it would take them to complete the spreadsheet testing, and they quoted an estimate of one year. We then turned to QARun, and realized we could do it much quicker—in one month instead of one year. And that's how it turned out; we completed the testing using QARun in 30 days. That's a huge savings."

Grech has gone on to use QARun in other Projects, including the testing of a huge, and critical precious metals system that was being updated. He said, "Basically I had to replace the hardware, the operating system and update the database. Before I began using QARun, I had to make a business case for using it.

My estimate was that, without using QARun, it would take 2.5 years of work. With QARun, we did it in five months. "QARun ensures consistency of input and ensures the output is based only on data from the system. It is a powerful tool, that is easy to learn, and easy to use. For large-scale system testing, I cannot see a tool that could be applied more easily than QARun."

Case Study – Compuware / First Horizon Home Loans [57]

Industry
Loan Origination and Servicing

Customer
First Horizon Home Loans

Challenge
The system testing team at First Horizon Home Loans wanted to manage testing tasks with greater ease and efficiency. It also sought to improve communications with internal Clients and Development Teams about system problems.

Solution
First Horizon chose Track Record, a product for coordinating software development, testing and deployment tasks, and communicating their results; QARun for automating setup and execution of Test Scripts; and File AID/CS, a comprehensive test data management tool.

Tools
QACenter

File-AID/CS

Background
As one of the largest mortgage companies in the U.S., First Horizon Home Loans services nearly 500,000 homebuyers. The company's steady growth has meant that its Information Technology department must manage and develop a growing variety of systems for loan origination and delivery, financial, internal work flow and imaging systems.

Testing these systems before they're put into production is the task of the Quality Assurance (QA) testing team. As a group that's independent of the system Development Teams within the IT department, the QA testing team acts as an objective interface. The team identifies business Clients' requirements for each system, objectively tests the system, and communicates the existence of defects and their resolution to both the Development Team and the Client.

> "TrackRecord makes for a much smoother testing and Implementation process. It's a win-win situation for everyone."
> **Carolyn Mattox, Vice President, Business and Process Management, First Horizon Home Loans**

With a strict business focus on its testing process, the QA testing team sought out tools to help it perform tests more quickly and accurately, and manage the entire process of testing, Documentation and communication. That's why the team turned to Compuware for many of its testing tools - TrackRecord in particular.

---

[57] Reprinted with permission from Compuware

Tracking Down Better Communications

"TrackRecord helps us track defects in systems very efficiently, and communicate in more detail with the developers," says Carolyn Mattox, vice president of business and process management in First Horizon's IT department. "The QA testing team manages the system defects it finds in testing. It logs each defect found into TrackRecord. The team can then tell the developers exactly what needs to be fixed. It helps the team provide excellent information to internal Clients as well; where the problems are, what needs to be fixed, and so on."

As part of the QACenter family of system testing products from Compuware, TrackRecord helps organizations establish and maintain a systematic method for tracking defects, from detection through resolution and verification. By handling labor-intensive tasks of documenting and reporting defects, TrackRecord helps organizations resolve problems quickly and improve system quality.

"TrackRecord has really saved us, because without it, we'd be tracking all this information manually," says Stephanie Madkins, testing manager. "Instead of the Test Managers spending all their time tracking information, they're actually spending time overseeing the testing process and working with Clients. Further, with the customized Reports from TrackRecord, we can really improve Client relationships, and that's important."

The QA testing team also uses other testing products from Compuware, including QARun for functional test automation. "It's very easy to use," says Mattox. "Even on small system test Projects, it makes creating the test and analyzing the results more efficient." The team also uses File-AID/CS, a comprehensive test data management tool to develop, test and support Client/server systems.

Staying Focused on the System

The QA testing team chose TrackRecord, in part, because the tool was easy to use. "We use TrackRecord daily now," says Madkins. "If you're testing, you always have to monitor your defects, and TrackRecord helps us do that. For example, the Notification feature in TrackRecord helps us inform the developers that they have a defect to repair, and we know everyone's been notified."

It's this ability to improve communications between different groups - testers, developers, and Clients - that makes TrackRecord important to the success of First Horizon's formalized testing process.

"TrackRecord has given us the ability to speak the same language to our internal Clients and our Development Team," says Madkins. "When we're in a Testing Phase, it's sometimes hard to keep the developer focused on the system at hand. By the time we begin testing a system, the Developer may have spent many, many months on it. They're ready to move on. "With TrackRecord and QARun, we can now show them exactly what problems the Test Team is encountering. We can provide clear instructions for the Development Team as to the defect's character, what it's doing to the system's performance, and what they should look at. Now they get much more accurate information, and it helps them spend their time wisely, too."

Keeping the Customer Satisfied

Working effectively with internal Clients is just as important as it is with Developers. "With TrackRecord, we can tell our Clients, 'Here's what we found, here's the percentage of items fixed, here's the defects we've certified as fixed, and here's the outstanding items that need to be addressed," says Madkins. "That's an important aspect of our success."

The QA testing team reaps the benefits of TrackRecord when it completes a system test and turns the system over to the Client for acceptance testing. From a list of defects still being worked on, for example, the team can show Clients that the Development Team is addressing them. Based on that information, the team can help negotiate decisions between developers and the Client, such as whether to go ahead when the defects are minor, or wait until the Development Team fixes the remaining defects.

"We're committed to providing the finest service to our Clients, and TrackRecord goes a long way in helping us achieve that," says Mattox. "The more our Clients see us providing excellent service, the more confidence they have in their systems when we deliver them. TrackRecord makes for a much smoother testing and Implementation process. It's a win-win situation for everyone."

Ultimately, a powerful tool is only as good as one's ability to use it easily, and ask for service and support when questions arise. Here, too, Compuware delivers a complete package. "As far as service goes, Compuware really makes us feel important," says Mattox. "Other companies might sell you some software and never talk to you again, but not Compuware. We've had excellent, one-on-one service, better than I've seen from a lot of other companies."

Case Study – Compuware / Royal & SunAlliance [58]

Industry
Auto Insurance

Challenge
Royal and SunAlliance takes pains to provide Clients simple, reliable online transactions for its nonstandard auto insurance products. To do so, the company needed to increase its system testing scenarios without increasing staff or the workload of its current staff. The company needed an automated testing solution.

Solution
Integrating QARun and File-AID/CS proved to be an effective solution. The tools saved the company time and resources, while offering more consistency, flexibility and reliability.

Tools
QACenter, including:
- QARun
- QADirector
- File-AID/CS

Background
In the United States, Royal & SunAlliance offers a nonstandard auto insurance product marketed under the name OrionAuto, which covers high-risk drivers. The nonstandard operation is known for its quick claims handling and competent Client service, which is reflected in its tagline, "Insurance. Easy to Start. Easy to Stay," and in the careful measures it takes to maintain Client loyalty.

> "I'm continually amazed at the potential for using these tools. I had no idea a year ago that I would be able to do some of the programming things I'm now able to do. The tools helped me learn. We got more than we expected from these tools."
> **Susan Donker, Quality Assurance Analyst, Royal & SunAlliance**

Providing simple, reliable methods for transacting business is critical to those efforts. That's why the QA team pays meticulous attention to testing systems that pertain to policyholder information. Royal & SunAlliance found Compuware's QARun, QADirector and File-AID/CS to be the perfect solutions to its OrionAuto system testing and data management needs.

---

[58] Reprinted with permission of Compuware

One of the company's main agent systems, EZLink, allows approximately 13,000 independent agents in 37 states to transmit business data to the home office, electronically. AutoLink, another important business system, is a web-based service that allows agents to perform online business transactions, such as submitting new business actions, and making online premium payments and account inquiries. Compuware's tools are used to perform unit and component testing of the EZLink system and the AutoLink web site. Quality assurance analysts check for valid and invalid entries in all fields, as well as for expected program flow.

More Consistency, Reusability and Flexibility in Less Time

Rate revisions drive Royal & SunAlliance's system testing, which previously required a considerable amount of time and resources. "Our testing consisted of sitting in front of the terminal and typing in all of the data that would allow us to get a rate out of our system," says Susan Donker, quality assurance analyst at Royal & SunAlliance.

"Then we sorted the data into a spreadsheet and visually compared it to our baseline data. The process was extremely manual. Nothing was automated." In late 1999, as workloads became increasingly heavy, the company's QA team realized it was time for a change. "We wanted a tool that would reduce the amount of time our staff required to perform manual testing," says Donker. The company was also looking for a tool that would run tests consistently. "If I do a test in one state, I want to make sure I do that same test in every other state. Compuware tools give us that kind of consistency, across the board," says Donker.

Testing Tools Offer Insurance Company More Security, Time, Resources

As new players continue to enter their market, insurance companies struggle to compete for Clients. Royal & SunAlliance realizes the secret to success depends on its ability to provide agents and Clients with reliable tools to perform online business processing efficiently.

In addition, it was important for Royal & SunAlliance to find a tool that was easily reusable, and that would interact with their mainframe, DOS, Delphi and Internet systems, and various Windows environments.

> "The products have allowed us to test thousands of scenarios in a relatively short amount of time. In some instances, we have been able to increase our testing thoroughness in addition to reducing some elapsed testing times."
>
> **Zach Osterloo, Technical Support Coordinator, Royal & SunAlliance**

Seamless Integration

Royal & SunAlliance uses Compuware's products to create multiple fields of random data every day, based on business requirements. With QACenter, the IT team was able to automate the entry of this data into the mainframe and into a spreadsheet. "File-AID/CS' Converter function provides us with translated insurance coverage cost data," says Zach Osterloo, technical support coordinator. The Compare function then lets the technical support team easily compare two sets of costs.

"The way all of the products tie together is very important to us," says Donker. "Everything integrated together is just a great fit for our company."

Ease of Use

Royal & SunAlliance uses QARun and QADirector to check system formatting. "QADirector helps us schedule tests," says Donker. "Rather than walking through the QARun step, then the Converter step, then the Compare step, QADirector handles the entire process with the touch of a button. And QADirector can automatically schedule these tests for a future time frame."

The company's technical support team is impressed with the ease of use of QACenter and File-AID/CS products. "We especially enjoy the functionality of the 'record' command in the QACenter products," says Osterloo. "It offers a great way for us to develop code quickly." Osterloo also commented that the language was very easy to learn. Employees with little formal training easily performed day-to-day development duties.

What is the overall benefit that Royal & SunAlliance has received from QACenter and File-AID/CS products? It's the company's ability to increase testing scenarios without increasing staff. "The products have allowed us to test thousands of scenarios in a relatively short amount of time," says Osterloo. "In some instances, we have been able to increase our testing thoroughness in addition to reducing some elapsed testing times."

"I'm continually amazed at the potential for using these tools," says Donker. "I had no idea a year ago that I would be able to do some of the programming things I'm now able to do. The tools helped me learn. We got more than we expected from these tools."

Royal & SunAlliance is convinced its Implementation of Compuware products will greatly improve the future of its quality assurance testing. Compuware makes it easy for organizations to start - with the development, integration, testing and management of their systems. And when the tools save them time and resources, it's easy for Clients to stay with Compuware.

Case Study – Mercury / John Hancock Financial Services [59]

Industry
Provider of insurance and investment products and services to retail and institutional Clients

Challenge
How, in an incredibly competitive marketplace, does a leader maintain and further its brand, better its bottom line and offer its Clients critical Web-based services that are dependent on a highly complex distributed IT infrastructure?

Solution
John Hancock Financial Services (JHFS) uses LoadRunner, Mercury Interactive's load testing tool, to load test all new Web-based systems prior to rolling them out to Clients. In addition, LoadRunner helps the company ensure that system upgrades perform problem-free and at optimal levels, and that existing systems are fine-tuned post deployment.

Testing Highlights
- Web-based systems, such as purchasing systems and a recruiting portal, are outsourced to System Service Providers (ASPs). JHFS uses LoadRunner to ensure that service level agreements (SLAs) are met by load testing systems to ensure that they scale to specified levels.
- Customers are provided robust Web-based self-serve systems, challenging JHFS with the task of continually upgrading existing systems without inconveniencing Clients. LoadRunner ensures upgraded systems scale and can handle thousands of concurrent users with ease.
- At any given time, JHFS has more than 50 critical Client-facing systems running in a highly complex distributed environment. The company conducts ongoing load testing of existing systems to ensure that systems are running at peak performance and offering the very best Client satisfaction.

Benefits
- JHFS believes that fast, reliable systems are vital for a source of revenue, increased business and increased Client satisfaction. Load testing has greatly reduced the risk of post-Implementation issues that could lead to dissatisfied Clients.

---

[59] Reprinted with permission from Mercury Interactive Corporation

- Load testing has led to improved efficiencies and a greater ability to serve more Clients, adding to the bottom line.
- A rigorous load-testing regimen has played a role in the national recognition and awards for excellence the JHFS Web site has received.
- JHFS Clients are benefiting from improved response times and increased performance from well-tested systems.

LoadRunner Pushes JHFS Systems to the Limit

LoadRunner exercises an entire enterprise infrastructure by emulating thousands of users and employs performance monitors to identify and isolate problems. By using LoadRunner, JHFS can minimize testing cycles, optimize performance and accelerate deployment.

"The AP group deals with a wide array of systems that span Client/server, multi-tier and outsourced ASP, making it crucial that any tools employed are able to support a wide variety of protocols," noted Verma. "At JHFS it is key that we work with an industry-standard tool like LoadRunner that can meet all of our corporate load testing needs without further investments in time and expense."

Today, JHFS' load testing procedures are fully automated. However, prior to using LoadRunner, JHFS manually programmed scripts. The AP group manages and supports all the performance needs for current and new systems at JHFS, including upgrades of existing systems and reactive support for systems already in production. Verma explained, "Our group provides Load and Stress Testing services, using LoadRunner, to our Clients, as well as database (SQL) tuning services, and continually works with our Clients until their performance needs are satisfied."

JHFS has been relying on LoadRunner for more than seven years, and with the growing use of the

> "Our primary concern is to verify that our hardware infrastructure is robust and scalable for supporting new Applications and a high number of concurrent users. Using LoadRunner, we are able to identify performance bottlenecks and enhance the performance of Applications, meeting the SLAs of our Clients. We've found that LoadRunner Reports are accepted as an industry standard for meeting performance targets."
>
> **Mohit Verma Consultant, Application Performance Group**

Internet for doing business, the company's Web-based architecture has grown in popularity to serve Clients and expand the business. Due to the inherent distributed infrastructure of modern computing systems, the complexity of systems has increased, and performance engineering has become more difficult and important.

"JHFS has incrementally added more Web-based services to its Clientele, and most new Projects are working in a distributed environment—Net-centric or traditional Client/server," said Verma. "As for changing and upgrading systems, performance testing is critical to verifying that the system response times are within the specified business requirements - if not improved - after upgrades or changes."

LoadRunner Verifies That Service Levels Are Met

One key factor in JHFS' successful use of LoadRunner is the tool's ability to help managers ensure that service level agreements (SLAs) are enforced. Another is that the product easily supports the vast set of standards and protocols inherent in the company's complex distributed architecture. JHFS' primary goal with load testing is to verify that its hardware infrastructure is robust and scalable for supporting new systems and a high number of concurrent users. Often, the AP group will determine software or hardware upgrades as identified by LoadRunner.

According to Verma, several systems are outsourced at ASPs, and load testing is essential to verify that SLAs are being met. "For example, sourcing and purchasing systems and a recruitment portal are hosted at an ASP with a strict SLA.

Stress testing allows us to identify performance bottlenecks and enhance the performance of systems, meeting the SLAs our Clients require. We find that LoadRunner Reports are accepted as an industry for meeting performance targets."

One system at JHFS with a strict SLA is a critical long-term care system. Verma's team is required to keep the system running at a peak performance of up to 1,000 concurrent users. In initial load tests, his team found that the system responded poorly at just 400 concurrent users. "LoadRunner Reporting and monitoring results helped us to identify the key bottleneck, which, in this case, resided at the system server level. After correcting the problems, we re-ran our load tests and were able to scale to the 1,000 concurrent user level—much to the delight of our Clients."

JHFS has found that LoadRunner supports the widest array of systems, from Client/server to multi-tier, with protocols such as DBlib, (MS SQLServer) ODBC, HTTP, (Oracle) OCI/UPI, CORBA and more. JHFS has had great success using LoadRunner to exercise the company's entire Web infrastructure: Client/server systems running on Windows NT and UNIX servers on the back-end; Windows NT Clients developed in various languages, such as Visual C++, Visual Basic, PowerBuilder and Java; and Web-based systems utilizing Java, XML and JavaScript. "LoadRunner meets all of our corporate load testing needs," stated Verma.

LoadRunner Improves the Bottom Line for JHFS

In addition, the AP group is able to use LoadRunner to troubleshoot and calculate capacity Planning for its business units. Most load testing at JHFS is conducted in production-like environments, and Verma and team work with the server, database and network monitoring teams so that these tests are fully monitored. The results have helped JHFS identify and correct issues with systems proactively, resulting in satisfied Clients and improved productivity.

Perhaps JHFS' biggest challenge has been its transition from a mutual life insurance company to a publicly traded insurance and financial services company, dependent in part on Client interaction transpiring on the Web. "The John Hancock brand is one of the most well recognized names in the financial services industry today," noted Verma. "The company distributes products and services through a wide range of channels and continually strives to improve the quality of its products and services. We see our role in the AP group as helping the entire company improve the bottom line."

Customers benefit from the improved response times and the increased performance of well-tested systems. According to Verma, testing improves Client service, boosts efficiency and allows JHFS to serve more Clients to their satisfaction. Moreover, testing systems reduces the risk of post-Implementation issues, which can cause the company to lose business. Fast, reliable systems are a vital source of revenue, increasing business and Client satisfaction.

Case Study – Mercury / NASDAQ [60]

Industry

Brokerage

Challenge

How can the largest U.S. stock exchange maintain the integrity of massive, constantly updated systems, accountable to member firms, the investing public and the SEC, while handling $17.7 billion in trades daily?

---

[60] Reprinted with permission from Mercury Interactive Corporation

Solution

TestDirector Open Test Architecture as the core of WinRunner and LoadRunner testing across all NASD systems and platforms

Test Highlights

- Single Test Management frame-work for 70+ NASD systems
- Unified control over functional and Stress Testing across diverse systems and platforms
- Single platform for test creation, execution and data storage and single Test Case repository for future reuse
- Enhanced, long-term efficiency for functional testing
- Future Plans include Web Test Management for one of the largest intranets in the world, with sites that register up to 12 million hits per day

Benefits

- Effective production systems at deployment, fewer releases over long term
- Lower development costs, optimum use of personnel and resources

High-Profile Success, Strict Monitoring Requirements

Operated by the NASDAQ-Amex Market Group, The NASDAQ Stock Market is the largest electronic stock market in the world, trading more shares per day than all other U.S. exchanges combined. In addition, its parent company, The NASD, which supports and monitors NASDAQ activities, is one of the largest self-regulated organizations (SRO) in the world. It has approximately 5,500 corporate members, with about 550,000 End Users who regularly access The NASD's online Applications.

At present, NASD systems handle up to $17.7 billion in trades per day, which translates into approximately 647 million shares daily. Trades for 1997 reached $4.5 trillion. The organization's complex Applications, residing on diverse platforms, expand by 30–50 GB per day, with 1.3 terabytes kept online for easy access.

Further intensifying the demands on The NASD's information services are strict reporting requirements from the U.S. Securities and Exchange Commission. The NASD is obligated to ensure that all member firms "play by the same rules," and that the investing public is treated fairly by brokers. In fact, The NASD's monitoring responsibilities are so comprehensive that any breach in control could result in a shutdown.

New Efficiency and Control

To improve efficiency and accuracy, and to reduce development costs in this massive, diversified environment, The NASD has begun a long-term program to revise its enterprise-wide Application-testing environment. TestDirector was selected to provide the overall testing methodology and architecture in which all testing will take place. "Our Test Management group has become the largest department in IT, but size is not always better," says Eric Henry, NASD associate director of Test Management. "TestDirector has given us a way to provide consistency throughout the organization, regardless of the Application being tested, the kind of test being run, or the testing mode - manual or automated - we choose. The result is tighter control, better access to test information and a lot more reuse of test information over time."

> "If something were to go wrong, so that we couldn't monitor trading, the SEC could shut NASDAQ down. For us, mistakes just aren't an option. TestDirector gives us the kind of control we need to test our systems repeatedly in an efficient manner, so that trading can continue to proceed smoothly, fairly and quickly."
> *Eric Henry, NASD Associate Director of Test Management*

PeopleSoft First in Queue

The first Application to be tested with the new TestDirector methodology was an upgrade of PeopleSoft financial software. TestDirector Backs Integrity of $4 Trillion in NASDAQ Trades Annually. Henry said that "the upgrade required lots of customization and data compares, with many repetitive functions that could be easily tested with WinRunner. This was a quick victory for WinRunner, which is perfectly suited to the iterative kind of testing we needed."

PeopleSoft testing included approximately 150 tests run over four months. "TestDirector made an enormous difference here," says Henry. "For instance, it allowed us to execute an entire suite of WinRunner tests using a single TestDirector File. Also, we converted a number of manual tests into automated format, which TestDirector managed beautifully."

Henry adds that TestDirector can even be brought in "late in the game" during development without excessive overhead and quickly reveals the advantages of automated testing for appropriate Applications.

TestDirector Features Save Time and Effort

Henry points to a number of specific TestDirector capabilities that have made his job a lot easier, and have resulted in greater efficiency during testing:

Identifies Functions Being Tested

TestDirector can specify and record the particular functionality being tested by each WinRunner script. When the same Application is later revised, the Test Team can select scripts according to the functionality that needs to be re-tested, rather than having to run an entire test suite.

Addresses Special Situations

Proprietary software is another candidate for manual testing, since automated tests often require hours of special code development to accommodate proprietary features. Here also, TestDirector allows the team to utilize manual methods in order to save valuable time.

Provides "Soup to Nuts" Functionality

TestDirector provides a centralized environment for designing, creating, and executing tests, as well as storing the resulting data. Project data can also be transferred to an Oracle repository for future reference.

Maximizes Enterprise Resources

Scripts can easily be reused, and information gained during one set of tests can be utilized as needed by other Test Teams, even with different platforms or Applications.

Web Testing to be Enhanced with TestDirector

The NASD has already started Web testing with WinRunner to certify a few of its Web sites for Year 2000 compatibility. However, Web testing will become even more critical as The NASD's legacy Applications are moved to thin-client, Web-based environments. Henry says that the volatility of these sites makes automated testing essential. "We have one of the largest intranets in the world," Henry says. "Nasdaq-Amex.com alone takes an average of 12 million hits per day, and has gone as high as 20 million hits in a single day. At that level, a Web Application can develop problems very quickly if it's not monitored carefully." Apart from the volume of Web activity, there is also the fact that Web sites change rapidly with multiple links being added daily. "This is where TestDirector and LoadRunner will really come into play," Henry adds. "They have given us a dependable framework for monitoring our Web Applications and continually subjecting them to functional and load testing." Established Vendor, Long-Term Strategies Henry also mentions that one of the major factors leading The NASD to select TestDirector was the stability of the vendor itself. "We felt that Mercury Interactive was the established leader in its industry, and had a big investment in keeping its Clients happy. We've also found that the company has a strong sales and technical support infrastructure, providing the quantity of people necessary to handle Clients properly, as well as the expertise required to answer our technical questions."

For Henry, this ensures a long-term relationship that will allow The NASD to extend its testing strategy well into the future. TestDirector has helped lay the foundation for technical growth and change in an environment of confidence, giving The NASD a dependable paradigm for success over the long-term.

Case Study – Rational / Industrial & Financial Systems [61]

Industry

Financial Services Software

Challenge

Industrial & Financial Systems (IFS) wanted to load test their component-based Applications with realistic workloads to establish accurate sizing Guidelines. They needed a scalable, reliable and easy to use performance-testing tool.

Solution

Use Rational Suite TestStudio to test the performance characteristics of the Applications.

Tools

Rational Suite TestStudio

Benefits

- Increased development speed by quickly evaluating different architectural approaches with realistic tests
- Improved software quality and performance by pinpointing bottlenecks and other potential problems during Load Tests
- Rapidly evaluated product performance on a number of platforms at one of IBM's Solution Partnership Centers
- Quickly created comprehensive, realistic load testing scenarios by recording Test Procedures, graphically creating workload schedules, and scaling tests to thousands of users
- Established accurate sizing Guidelines, which are used by sales personnel to recommend appropriate hardware for any IFS Application and expected user load

Vendor B, or not Vendor B

Imagine for a moment that your organization needs new enterprise-wide business software, and you are leading the evaluation effort to recommend a solution. You have narrowed the choices to two vendors. Both Vendor A and Vendor B offer similar features at a similar price. When you ask Vendor A how many users their software can support on your preferred platform, they wave their hands and change the subject. When you ask Vendor B the same question, they give you actual numbers of verified transactions and users that the software can handle on your preferred platform and several others - all backed with realistic, reliable, and repeatable test data. Which vendor would you choose? The answer is obvious.

If your Clients need to know how your Applications will perform before they deploy them, then you share an urgent business need with IFS (Industrial & Financial Systems), one of the fastest-growing companies in the business Applications market. Based in Sweden, IFS develops and supplies business Applications that offer step-by-step evolution

---

[61] Reprinted with permission of the Rational Software Corporation

to the extended enterprise. With over 3,600 employees, IFS maintains a strong presence in 42 countries through 64 offices around the world. IFS uses Rational Suite TestStudio (formerly Rational Suite PerformanceStudio) to load test its software, enabling them to accurately predict how their key Applications will perform across a wide array of available Applications. Per Bauer, Benchmarking Manager for Research and Development division at IFS, reports that Rational Suite TestStudio has been a big part of IFS' success. "We've been able to beat our development schedule on a number of occasions, because Rational Suite TestStudio helped us evaluate different architectural approaches with realistic tests. This gives us a real competitive advantage."

Rational Suite TestStudio makes it easy to create load tests involving tens of thousands of users and millions of individual transactions. In a recent test, Rational Suite TestStudio applied a load of 7,000 concurrent virtual users to IFS software. The 7,000 virtual users - a number well within Rational Suite TestStudio's capabilities - performed a variety of business activities while Bauer and his team used Rational Suite TestStudio to monitor Application performance and gather performance measurements. Bauer was very happy with the overall results, "From a Rational Suite TestStudio point of view, the tests were most successful. During the whole period, we did not have any problems at all. Our growth and success depends, in part, on our ability to provide our Clients with accurate performance Guidelines, and we rely on Rational Suite TestStudio to get them."

The Business Problem

Before IFS began using Rational Suite TestStudio they faced a common problem - they needed to load test their Application with realistic workloads. Bauer, who had been with IFS for two years at that point, remembers, "We needed a way to test our Application with extremely high realism and we needed to be able to establish Guidelines for the Application. We evaluated a couple of other testing products at the time, but we were already using other Rational tools, like Rational Rose, and we were very pleased with their quality."

Since then, Bauer has used Rational Suite TestStudio extensively to test IFS Applications - IFS' flagship product. Based on proven component technology, IFS Applications include Web-based enterprise Applications, supply chain management Components, Internet storefronts, and e-markets and e-procurement solutions. The company offers over 60 functional business Components - spanning the entire demand and supply chain - to improve business processes in medium to large companies. IFS eBusiness solutions provide seamless integration of all data sources from the fastest Web sites and exchanges to point solutions and legacy ERP (Enterprise Resource Planning).

IFS Applications include Components that span the entire demand and supply chain, including Web and portal-based solutions as well as Internet-based business models for sales, marketing, procurement, and dynamic collaboration.

Bauer and his team of six test engineers are responsible for establishing sizing Guidelines for IFS Applications - including the more than 60 business Modules that can be assembled in any combination to meet Client requirements. These sizing Guidelines help IFS Clients determine an appropriate Application configuration for their particular needs. IFS has been using Rational Suite TestStudio for almost three years to help determine the maximum transaction rate and maximum concurrent user load that their software can handle.

Bauer explains, "With our interactive sizing guide, a Client can tell us how many users they have and what software they intend to run, and we can suggest a suitable platform. There are also many unique Client cases that fall outside our Standard Guidelines, so we do a lot of presales and individual sizing Projects for special Client cases. Our Clients require this information to make informed decisions, and of course, it has to be accurate. Accurate sizing Guidelines help keep our Clients' costs to a minimum."

Preparing for The Big Test

The recent 7,000-plus virtual user test was one of the more extensive test scenarios that Bauer has conducted at IFS. The tests took place at an IBM Solution Partnership Center (SPC) in Waltham, Massachusetts.

Bauer Reports, "Our SPC tests were basically aimed at establishing Guidelines for IBM RS/6000 servers and IBM environments, like AIX for example. We had two Applications that we used interchangeably as the Application under test and the load server.

We were able to perform tests on AIX using TestStudio's UNIX load test agents, which worked very well." Rational Suite TestStudio supports performance testing in a wide range of environments with playback agents on a variety of platforms – including IBM AIX, Sun Solaris, HP UX, Red Hat Linux, and Windows NT. Bauer

> "It is a very straight forward process. Rational Suite TestStudio is highly effective in this regard and it is also very stable during testing. When we update a script, we just record new segments for the added features. By reusing scripts this way, we save time and shorten the entire testing and development cycle."
> *Per Bauer, Benchmarking Manager for Research and Development Division at IFS*

continues, "We used an IBM Netfinity server as a load master for report processing and other tasks."

Bauer's team prepared for the tests at IFS by recording scripts and setting up the test environment. They then brought the scripts and their test environment with them to the IBM Solution Partnership Center. Bauer continues, "Our team was at the SPC for about ten days, and we ran more than 7,000 virtual users on the Application. That is very good because those are realistic users performing a lot of activities on the IFS Application. We focused on those Components that are normally high-volume in a Client installation because those Components tend to determine the sizing of the required hardware. So in the test we processed a lot of sales orders, did a lot of inventory procurement, and a lot of financing and invoicing."

With Rational Suite TestStudio, Bauer's team is able to prepare for even the most extensive tests with relative ease. Using the Rational Robot component of Rational Suite TestStudio, Bauer quickly records new Test Procedures whenever a new feature is added to an IFS Module. During the three years that IFS has been using Rational Suite TestStudio, they have assembled an extensive collection of test scenarios. "We upgrade our scripts when we release new IFS Application versions. We extend the scenarios a little, modify them, or add new ones if needed," Bauer continues. "It is a very straightforward process. Rational Suite TestStudio is highly effective in this regard and it is also very stable during testing. When we update a script, we just record new segments for the added features. By reusing scripts this way, we save time and shorten the entire testing and development cycle."

From One Virtual User to Many

Typically, a recorded Test Procedure will perform one or more user activities, like placing an order for example. For that script to be used effectively in a performance test, the script must use a range of data (or variable data) when it is played back. Consider a script that added a new part number to a database. During

> "The ability to scale tests to very large numbers while maintaining realism is a very powerful feature – and it is something we need to continue to meet our Client's needs."

a load test that script is executed many times, and a virtual user would try to add the same part number each time. The results of such a test would likely be of little value, since attempts to add the part number after the first would likely be rejected by the database. Rational Suite TestStudio uses data pools to create multi-user scripts that use a range of values – or a set of enumerated values – during playback. When a multi-user script is used in a performance test, Rational Suite TestStudio automatically varies the data used in the script based on the test engineer's specifications. Bauer confirms that creating a multi-user script from a single-user script with Rational Suite TestStudio is simple, "The quality of the scripts you get from recording sessions is very high, and so you can rapidly add more functionality. For example, we can easily script different behavior and add data pools. We use data pools to generate different data and, to some degree, cause random behavior during our tests. Setting up a data pool with Rational Suite TestStudio is intuitive. We use them to vary everything from Client data, to order data, and part numbers."

Ramping Up

In the real world, there are many different kinds of users using a software system in different ways. For example, while Clients place orders, managers check inventory, and accounting personnel create invoices. Effective performance tests must simulate the activities of each of these different groups of users, giving each group a proportionate amount of work to do. Rational Suite TestStudio enables testers to quickly build complex usage scenarios without Programming. Testers create very realistic workload schedules by weighting each user group. For instance, a simple schedule might specify 50% of users as Clients, 20% as report-generating management, and so on.

Once a schedule is created graphically, the whole test can be easily scaled from ten users to thousands of users.

With this capability, the IFS team had a high-level of confidence that the benchmark workload they created was representative of the real world. Then they simply ramped up the number of users in the test. Bauer recalls, "We had a set of approximately ten different roles on the Application – end uses pertaining to sales, procurement and the financial department and so on. We assigned a percentage to each group, and when we scaled up we got the same distribution among users. For each role there were four or five different activities and intensities inside each role. Some activities were done once every five minutes, some were done once every fifteen minutes. The ability to scale tests to very large numbers while maintaining realism is a very powerful feature - and it is something we need to continue to meet our Client's needs."

Rational Suite TestStudio simplifies the process of creating accurate, scalable performance tests that employ thousands of virtual users to carry out millions of individual transactions. It also gives Testers, including Bauer's team, the ability to precisely control the rate of transactions. In Rational Suite TestStudio transactors are used to set the number of Tasks that each virtual Tester will run in a given time period. Bauer continues, "We use transactors a lot. We can set a pace for a user and we do not have to worry about how long it actually takes to carry out the scenario. Rational Suite TestStudio takes care of the pacing of user activities. If we want something done twelve times an hour, we can set that and we know TestStudio will do it."

Getting Results

The ten days Bauer and his team were at the IBM Solution Partnership Center proved to be time well spent. IFS came away with not only sizing Guidelines, but also valuable insights into key performance issues. Bauer remembers, "One of the benefits of going to the SPC is you have a wide selection of different servers you can test. Normally, during the first part of the test, we focus on tuning the hardware and the Application software to avoid bottlenecks there. We were able to tune our Oracle Database for these high-volumes once we found out where contention issues were. Also we were able to identify some network bandwidth issues."

Rational Suite TestStudio also helped highlight specific areas of Code in IFS Components where changes might significantly improve performance. Bauer continues, "But when you get past hardware-tuning levels you start to find out things inside the Application - Implementation details - that are not 100% effective. We found some of those and were able to fix them during our tests. When we did find an issue, Rational Suite TestStudio helped us pinpoint exactly where in the source Code the problem was. As a result, we were able to improve the reliability of our Applications."

With Rational Suite TestStudio, Bauer analyzed everything from Application resources to business transactions to low-level function calls. "With a very large-scale test like this one, when everything goes right, you look at it from a high level and only measure large blocks of transactions to see how long they take. But Rational Suite TestStudio also has a set of very fine-grained timers on each action you perform. So when you see things starting to take more time and become slower, TestStudio can actually drill down inside them so we can see how different transactions are performing at a detailed level. That is very useful for us. When we find potential areas of improvement we give feedback to our Developers. We can't always suggest how to improve something, but at least we can show what areas we would like to see improved," Bauer notes.

Improved Speed, Improved Quality, Improved Business

IFS, like all successful software Development companies, must develop software faster than ever before, while constantly improving on their already high-quality standards. Bauer believes that Rational Suite TestStudio has helped IFS do just that, "To start with, Rational Suite TestStudio has enabled us to find out how our Application behaves under massive loads and high pressure. That is basically the most important part of it.

Simulating an Application with hundreds or thousands of users in the lab with 100% realism - you can't do without that. It's a must in all cases. In that sense, it helped us a lot in increasing quality. Our speed has improved because we save time in creating and maintaining our tests with TestStudio. And we can better serve our Clients because we are able to provide reliable, accurate information on what our Applications can handle on a variety of platforms."

For Bauer, the key benefits of Rational Suite TestStudio are the realism of its tests, the ability to develop tests rapidly, scalability, and accuracy. Although confirming the accuracy of load Test Results with real world data is not easy, Bauer has managed to do it. "We have been able to verify the accuracy of our tests. Because we have an Application that covers a wide range of functions, it is pretty rare that a Client will run the same configuration as we did during our tests.

However, it has happened, and we have found for a given platform if an Application doesn't scale past a certain point during testing it doesn't in reality either. Also, we get feedback on all our sizing Guidelines. With Rational Suite TestStudio, we find that the feedback is in line with our tests, so we are confident in our ability to predict how our software performs in the real world. When we can provide this kind of assurance to potential Clients, it really helps our business."

More Than Just Testing

Before Bauer and his team began using Rational Suite TestStudio, IFS was already taking full advantage of other Rational products, including Rational Rose. Developers at IFS have been using Rational Rose, the world's leading visual modeling tool, since 1996 to model business objects, attributes and methods. IFS Developers also use Rose to generate Code for their PL/SQLserver and Windows Clients directly from their visual models. IFS is currently considering using its Rational Rose models to create component tests. With Rational QualityArchitect, Testers can quickly and easily generate functional tests for Components modeled in Rational Rose.

In addition to Rational tools, IFS also leveraged Rational e-Development Services to streamline the deployment and adoption of Rational Suite TestStudio. Bauer continues, "We had a proof-of-concept period where we Implemented Test Procedures for a part of our Application suite. Two consultants from Rational helped us for about a week and after that we decided on Rational. When we started using Rational Suite TestStudio, we attended Rational University courses at our site, which definitely accelerated our deployment. We got a kick-start, and started using TestStudio immediately after that. Now, after three years with TestStudio, the biggest challenge for us is figuring what to test

- not how to test. It's easy to do the tests now, the hardest part is coming up with tests that represent all the different ways our Clients use our product."

Imagine That

Clearly, IFS has been very successful with Rational Suite TestStudio over the years. With its ease-of-use, scalability and accuracy, it is little wonder that Bauer has put Rational Suite TestStudio's load testing capabilities to use wherever possible. He explains, "Rational Suite TestStudio has grown to become our primary tool for all kinds of benchmarking and performance improvements. We are using it for both our Web-based and our traditional GUI Applications. We are also using Rational Suite TestStudio to test part of our Application built on a CORBA platform. We are using it for almost everything we are doing."

Even after using Rational Suite TestStudio on many Projects, Bauer continues to be amazed at its usefulness and its value. "If I look back to before we started using Rational Suite TestStudio, I never would have imagined that a test tool could be this powerful — that you could get high level of realism, and really run one, two, five-thousand users and have them behave like five thousand real users. With TestStudio, you can really do this and that is very important for us. I am still really impressed that it can be done. And on top of that, it is easy. It is easy to orchestrate the whole test, and everything will happen exactly as you want it to."

Now, imagine for a moment that your organization needs a new load testing solution, and you are leading the evaluation effort to recommend a solution. If, like IFS, you demand effective, scalable, and easy-to-use tools, then your clear choice is Rational Suite TestStudio.

Appendix VI - Master Test Plan Checklists

We have presented three sample checklists in this Appendix that are designed to help with:

- Overall Test Management
- Writing the Master Test Plan
- Performing Functional Testing.

Test Management Checklist

Topic	Category	Sub-Category
Management Process	Identify Management Guidelines	Define Code Delivery Criteria
		Vendor Service Level Agreement
		Define Entrance/Exit Criteria
		Define Suspension/Restart Criteria
		Define Defect Turnaround Criteria
	Identify Defect Management Guidelines	Define Defect Process
	Develop Standards/Guidelines	Test Methodology
		Test Standards
		Documentation Standards
	Finalize Contacts/Agreements	Outsourcing
		Vendors
		Service Level Agreements
		Hardware
		Software
	Staff the Team	Employees
		Consultants
	Assess Need for Automation	
	Develop Procedures	Issues Tracking
		Test Execution
		Test Verification
		Meetings
		Communications
MIS	Define MIS Parameters	Define Reporting Procedures, Method & Frequency
	Develop MIS Processes	Task Tracking
		Quality Tracking
		Scope Tracking
		Costs Tracking
		Risk Tracking
		Communications
Sign-Off Requirements	Identify Sign-Off Levels	Signoff Test
	Define Exit Criteria for the Functional Test Cycles	
Test Development	Test Preparation	Assign Test Case Preparation Responsibilities
		Set Up Accounts, etc...
	Prepare Functional Test [62]	Log & Track Defects

---

[62] We are only providing an example for the Functional Test.

	Develop Test Plans/Test Material	Summary Test Plan
		Detailed Test Plan
		Project Plan
		Business Functions
		Business Scenarios
		Test Cases
		Test Procedures
	Develop Test Criteria	Entrance/Exit
		Sign-offs Required
		Suspend/Restart
		Defect Turnaround Times
	Establish Test Environment	Setup Test Environment (server)
		Setup Test PC's
		Setup Issues Tracking Database
		Verify Test Environment
		Setup Test Data
		Setup & Install Test Peripherals (card readers, receipt printers)
Budget	Prepare Budget	
Close Project	Review Test Project	
	Produce Post Test Report	
	Cleanup Test Environment	
	Return Equipment	
	Reassign Staff	

Figure 182 – Test Management Checklist

Master Test Plan Checklist

The following checklist can be used by the Test Manager to ensure that they have covered all of the main questions that are involved in preparing the Master Test Plan. This checklist is an extension of the outlines that were presented in Chapters VI, VII and VIII. The points that are noted in this table should be converted into Tasks for the Project Plan. [63]

TOPIC	CATEGORY	SUB-CATEGORY
Definition of the Project	Background	Category of Project
		Project Sponsor
		Purpose of the Project
	Objectives	Specific Business Objectives that the Project Should Accomplish
		Specific Functional Objectives that the Project Should Accomplish
	Documentation	Documentation used to write the Plans
Assumptions	Testing Roles and Responsibilities	Development, Test Team and End User Responsibilities
		Identification of Vendor Testing Responsibilities
		Identification of Responsibilities in testing Dependency Projects
	Standards	Define Testing Standards
		Identify which Corporate Guidelines will be used
		Identify Design Standards
		Define Documentation Standards
	Timeframes	Timing Constraints

---

[63] There is an additional level of detail to this checklist that is available from the authors.

Appendix VI – Master Test Plan Checklists

		Estimated Start Date
		Estimated End Date
	Conditions	Identify Business Conditions affecting the Test Cases
		Identify Technical Conditions affecting the Test Cases
	Delivery Requirements	Vendor Testing Prior to Delivery
		Testing Accomplished by Developers
		Code Delivery Parameters
		Target Date for Specific Code Deliveries
		Quality Guidelines for Accepting Code
	Suspension / Resumption Criteria	Criteria for Suspension/ Resumption
		Approval Levels Required for Suspension/Resumption
	Responsibilities	Define Test Team Responsibilities
	Limitation of Test Plan	Functional Coverage
		Test Types to Conduct
		Scope of Technical Testing
	Support	Development Team
		Business Analysts
		End Users
Test Scope	Define Module Test Scope	Define Testing Scope
		Define GUI Test Scope
		Define Report Test Scope
		Define Interface Test Scope
	Define Business Functional Scope	Functional Business Tests
		Technical Business Tests
	Define Systems Technical Scope	Functional Systems Tests
		Technical System Tests
Test Phases	Objectives	Identify Phase
	Environment	Specific Requirements for each Phase
	Exit Criteria	Procedures Covered
		Defects Allowed
Test Calendar	Develop Test Calendar	Develop Test Plan
		Develop Test Schedule
		Setup Test Execution Progress Tracking Database
		Archive Test Documentation
		Define Start and End Dates for Each Test Cycle
Dependency Projects	Identify Dependency Applications	Feeder Applications
		Destination Applications
		Interfaces
		External Networks
Databases and Environmental Plan	Region Requirements	Number of Test Regions Required
		Availability of Regions
	Interfaces with Other Applications	List Interfaces
		Scope of Interface Development
	Define Database Requirements	
	Parallel Test Requirements	Scope
		Requirements
		Timeframe
	Back-ups	Identify Routine Backup Requirements
		Identify potential Special Back-up Requirements
	Data Sources	Manufactured vs. Production Data

Human Resources Plan	Staffing		Define Tasks Required
			Assign People to Specific Tasks
			Assign Test Roles & Responsibilities
			Identify Skills, Knowledge and Capabilities Required
			Should People Have a Certification, such as PMI?
			Communication Skills Required
	Training		Testing Methods, Concepts, Procedures and Tools
			Software Training
			Process Training
			Project Management
	Organization		Identify the Types of Staff Required
			Identify the Organization Plan
			Identify the Number of Staff Required
			Identify Sources of Staffing
			Identify When the People Are Required
Automation Plan	Identify Automation Opportunities		Select Relevant Tools
Quality Assurance Plan	Identify QA Objectives		
	Select QA Methodology		
	Implement QA Methodology		
Other Resources Required	Support		Define Systems and Development Support
	Facilities Requirements		Premises
	Hardware		What hardware will be required
			Identify Technical Support Requirements
	Software		Support Software
			Test Software
			Identify Support Requirements
	Other		Identify sources for all Requirements
Risk Management			Evaluate Risks
			Identify Mitigations and Tasks Required to Implement
Communications	Define Communications Plan for the Test Project		Meetings
			Reports
			Contents
			Identify Communication Support and Who Will Provide It
Budget	Prepare Project Budget		• Staff • Hardware • Software • T&E • Other?
Review and Approvals	Identify Approval Requirements		Obtain sign-offs on the Master Test Plan
Project Plan			Convert Summary Test Plan and Detailed Test Plans Tasks in to Test Project Plan Tasks with Start and End Dates, Dependencies, etc.

Figure 183 – Master Test Plan Checklist

Test Management Planning Checklist

As the Detailed Test Plans are developed, there are several actions that must be taken to plan the management of the tests.

Functional Testing Execution Checklist

Each Detailed Test Plans will have its own Execution Checklist. This checklist will include the major points that are necessary to prepare, Execute and Report on the activities involved in the Functional Test.

TOPIC	CATEGORY	SUB-CATEGORY
Prepare	Set-up[64]	Accounts
		Securities
		Positions
		Code Files, Rule Sets, etc.
		Counterparties
		Foreign Exchange
		Factors
		Corporate Action Announcements
	Test Documentation	Business Scenarios
		Test Cases
		Test Procedures
	Data	Manufactured Data
		Production Data
	Define Expected Results	
	Execution Checklist	
Execute	Run Tests	
	Run Interfaces	
	Run Batches	
	Run Exception Processing	
	Run EOD Processing	
Validate	GUI	
	Reports	
	Files	
	Application Interfaces	
Document	Test Case Status	
	Issues Tracking	Defects
		Other
Control	MIS	
Retest		

Figure 184 – Functional Testing Execution Checklist

---

[64] One of the most important Tasks in preparing to test is ensuring that the set-up information is correct. If not, testing will be invalid and significant time will be lost.

Appendix VII - Glossary

Term	Definition
Acceptance Criteria	The Standards that must be met in order for the Application to be accepted and either: • Moved from one Test Phase to the next, or • Moved into Production The Standards should include measurable criteria that verify that the Application: • Does what is specified • Does what is needed • Meets contractual obligations
Acceptance Test	See User Acceptance Test
Ad Hoc Testing	Testing that is performed without a structured Plan, on an as-needed basis. Structured Testing
Application	An Application is a set of Programs or Modules that can complete a Task. An Application has input, processing and output capabilities. Program, Module, Task
ASQ	See Automated Software Quality
Audit Test	Audit Testing occurs initially when the Audit department verifies that: • The Test Plan is adequate and meets corporate Standards • The Test Plan was followed throughout the testing effort • The appropriate Controls have been built into the Application and the supporting process • The Test Plan was completed with acceptable results
Automated Software Quality	The use of automated tools to perform tests and ensure software quality.
Automated Testing	The use of automated software tools to test Applications with a minimum of manual intervention.
Back-up Testing	This test is used to determine if the Application can be successfully and routinely backed-up at any point in the processing cycle. The test must be performed at various points in the processing cycle, such as: • At the beginning of the processing cycle • At various critical points in the processing cycle • At the end of the processing cycle Recovery Testing, Contingency Testing
Black Box Test	This test is performed when the Tester does not know how the logic of each Program works. Testers focus on the Application's Functions. Functional Testing is an example of Black Box Testing. Functional Test
Boundary Scenario	A Boundary Scenario is developed to test that the Application performs correctly at the extremes of the range of acceptable data.
Bug	A Bug is a Defect in an Application that results from an inadequate specification or a Coding Defect.
Business Integration Test	Business Integration Testing is one of the final tests conducted by the End Users, often with help from the Test Team. It is performed to ensure that all of the Components that passed the Functional Test can in fact work together as part of a normal business process flow. This includes the Interfaces, logic, Databases, other networks, hardware, etc.
Business Scenario	A Business Scenario is based on the Functions that are defined in the Functional Requirements Document. Business Scenarios are either suggested by the End User or it may be Designed by the Tester to exercise the logical processing flow of the Application under normal business Conditions.
Calendar Day	A Calendar Day is an actual day on the calendar. Testing may occur on one or more Calendar Days.
Capture/Replay Tool	A test tool that records key strokes of Test Transactions as they are entered into the Application being tested. These keystrokes are stored and can be replayed to re-enter the transaction whenever it is necessary.
CAST	See Computer Aided Software Testing
Changes Only Testing	This is a technique whereby only the Functions that were Changed are tested. After this test has been performed, then the rest of the unchanged Functions undergo Regression Testing. Regression Test
Checkpoint	A Checkpoint is a point in time or in a process where a manager's progress towards a Plan can be measured. Plan, Milestone
Code	Code:

	• As a verb, means to write the instructions in a computer language that will tell a computer what actions it should take under specific circumstances • As a noun, is the set of written instructions that tells the computer what to do
Code Complete	This is the Milestone in the SDLC where all of the Code has been completed and all of the Functions defined in the Requirements Document have been Implemented. Some Code Defects may remain. Systems Development Life Cycle
Code Coverage Test	This test determines how much of the Code has been exercised, and identifies the areas of the Code that were not tested and which may require additional Test Cases.
Code Freeze	This occurs when the Project Manager decides to stop moving Code into the testing region. This can occur at various points during the testing and again when the final testing begins. A Code Freeze may also occur in Production during high volume or sensitive periods such as during year-end processing, Testing Region
Code-Set	A Code-Set is a list of allowable abbreviations or codes that are available to users, which describes activities of Functions that can be performed by the Application. This Code-Set may be unique to the Application being tested, or it might be used in other Applications. Rule-Set
Comparison Test	Testing can be performed to compare one Application to another similar Application in order to select between them.
Compatibility Test	This type of testing is performed to determine how well an Application will perform in a particular Environment. This is normally done prior to selecting the Application to be used. Interoperability Test
Compliance Testing	This type of testing is used to determine if the Application is Developed in accordance with Regulatory Requirements and the corporation's Standards and procedures. Conformance Test
Component	A software Component is section of Code for which a discrete specification is available.
Component Testing	See Unit Testing
Computer Aided Software Testing	
Concept Phase	The Concept Phase occurs at the beginning of the SDLC when the Project Team members prepare a preliminary product definition and an initial analysis of the potential solution(s). Systems Development Life Cycle
Conformance Test	This is a type of testing that determines if a specification or Code conforms to a specific Standard. Compliance Test
Contingency Test	In this type of testing, the Testers identify potential operational situations that could affect the company's ability to run the Applications (i.e., power outages, natural disasters, terrorism, etc.), and then run a test to determine if the Application can recover from these situations and still meet the business Requirements. Recovery Test, Back-up Test
Conversion Test	In Conversion Testing, the Testers verify that the Code that was created to automate the Conversion of data from one Database (or File) to another performs correctly.
Coverage Analysis	See Code Coverage
Critical Success Factors	Critical Success Factors are activities that must be performed successfully in order for the entire Project to be considered a success.
Data Validation	Verification of the test data to ensure that it meets the Expected Results.
Debug	To identify the Defects in an Application so that they can be repaired.
Defect	A Defect is: • An Error in the Code that delivers an incorrect response (or no response) • Correct Code that does not meet the business Requirements
Definition Phase	See Requirements Phase
Dependency	A dependency occurs in testing when one test cannot be performed until another test is successfully performed. A dependency may also occur when Development in a related Application is needed.
Dependency Project	A Dependency Project involves Changes to some Application that is affected by the Implementation of the new Application. These Changes must be made simultaneously and must be tested jointly.
Design Phase	The phase of the SDLC where the Applications analysts evaluate the Requirements and determine how the

	new Application will be built.
	Systems Development Life Cycle
Desk Checking	Desk Checking is a process that is performed by a Developer prior to a Unit Test in order to check their work (or another Developer's work) against the Business and Functional Requirements and the Design.
	Unit Testing
Disaster Recovery Test	A Disaster Recovery Test is used to determine if the Application, the supporting hardware, and the Workflows can be Implemented quickly in a separate location if a disaster occurs.
	Contingency Test
Documentation Review	A Documentation Review is often performed during User Acceptance Testing, along with Procedure Testing, with a goal to verify that the User Documentation is correct and useful.
	Procedure Test, User Acceptance Test
Dynamic Testing	When an Application is tested by running (Executing) the Application, it is called Dynamic Testing. The testing performed by Testers and End Users is Dynamic Testing.
	Static Testing
End-to-End Testing	See Business Integration Testing
Entry Criteria	Entry Criteria are those activities that must be successfully completed before starting another Task or milestone.
Error Handling Tests	Error Handling Testing is a form of negative testing that is Designed to determine if an Application can properly deal with exception Conditions, such as erroneous input. The Application must be able to correctly respond to Defects and to accept corrections in a manner that can be properly performed by the users in the time available.
	Negative Testing
Exhaustive Testing	This is a testing concept whereby the Testers create all possible combinations of input values and Conditions to test specific Functions.
Exit Criteria	Exit Criteria are those activities that must be successfully completed before a Task or milestone is considered complete.
Expected Results	When establishing a Test Case, the Tester must also define what they expect to see when the transaction is processed. Without this, there is no way to verify that the Applications successfully passed the test.
Exploratory Testing	This type of testing is often performed by Developers as an unstructured form of a Systems Integration Test or by Testers in order to 'get a feel' for the Application. It does not take the place of any structured testing.
Feature	A feature is a Function of an Application that is emphasized when the Application is marketed.
Finding	A Finding is what is found and Documented as a result of a test.
Functional Requirement	During the Functional Requirements Phase of the SDLC, a Document is produced that identifies what the users want the Application to do. It describes the Functions along with any other Guidelines that the users will require, such as input and output methodologies, Interfaces, processing logic and presentation requirements.
	Systems Development Life Cycle
Functional Specification	See Functional Requirement
Functional Test	The objective of Functional Testing is to ascertain that the Application's Functionality matches the Functions that have been defined in the Functional Requirements Document. This testing segments the Application into smaller operational units that can be tested independently before being combined for a complete Business Integration Test.
Goal	A Goal is a point where a manager or a firm wishes to be at a certain time in the future.
	Standard
Graphical User Interface	A Graphical User Interface (GUI), pronounced gooey, is an Interface between a person and a computer that helps the user input requests and see the data that has been recorded by the computer. A GUI is based upon a conceptual method of interaction that uses buttons, icons and graphical information rather than the lines of text that are typical of text-based Interfaces.
	Text-Based Interface, Cursor
GUI	See Graphical User Interface
Incremental Integration Test	Incremental Integration Testing is the sequential testing of Functions that are added by Developers during Systems Integration Testing, one after the other, as tests are successfully performed. Firms sometimes use this technique during Functional Testing.
	Real integration testing is performed during Systems Integration Testing and Business Integration Testing.
	Functional Test, Systems Integration Test, Business Integration Test

Information Technology	The Information Technology Department is responsible for developing, maintaining and running computer-based applications.
Inspection	See Walkthrough
Implementation Test	If an Application is in Production in one environment and is subsequently installed in another environment for live Processing the User should require an Implementation Test. This can involve any number of the Test Types that were discussed in Chapter III.
	This is not necessary when the Code is migrated from one region to another in the same Environment.
Integration Testing	See Systems Integration Testing and Business Integration Testing
Interface Testing	This is a test that ensures that data is correctly transmitted from Application to Application. Interface testing is first conducted during Unit Testing and System Integration Testing. It is also performed again during Functional Testing and Business Integration Testing.
	Interface testing can be more complex if the Applications operate in different Environments, on different platforms, in different locations or use different languages.
	Unit Test, Systems Integration Test, Functional Test, Business Integration Test
Interoperability Test	This is a test that is performed to verify that the software can work with the hardware, such as printers, scanners, etc. This is performed during Functional, Business Integration, User Acceptance and Parallel Testing.
	Compatibility Test, Functional Test, Business Integration Test, User Acceptance Test, Parallel Test
Issue	An Issue is a problem that is identified during testing. It may or may not be a Code Defect.
	Defect
Issues Tracking Report	The Issues Tracking Report is a list of all of the possible Defects that are documented as they occur. The Report is updated as the Issue is passed from the Testers to the Developers, as the Developers react, and as the item is Re-tested.
IT	See Information Technology
Load Test	A Load Test is a type of performance test that is Designed to determine how an Application performs when it is working at its limits. There are several types of load tests:
	Volume Test, Stress Test, Storage Test
Logical Day	A Logical Day is a day of Production using Manufactured Data A Logical Day is usually constructed around Business Scenarios that require more than one day to complete. For instance, a Trade is entered on one Calendar Day and does not always settle on the same Calendar Day.
	It differs from a Production Day in that the Production Day uses Production data that is related to an actual day of Production.
	Manufactured Data, Production Day, Calendar Day
Management Information System	MIS is a generic term that describes groups of information and Reports that are used by managers to: • Assess a Project's status • Facilitate remediation when required • Improve overall system and managerial performance
Metrics	A Metric is any measurement Standard that has been defined by a firm to monitor the Development, testing or operation of an Application.
Milestone	A Milestone is a measurable Task or a Checkpoint in a Plan that identifies specific events. Milestones can be organized into hierarchies with any number of levels, such as Major Milestones, Milestones, and Minor Milestones, etc.
	Checkpoint
MIS	See Management Information System
Model Office Testing	See Operational Test, Process Flow Test, Procedures Test, Simulated Work Environment, White Room Testing, Usability Test
Module	A Module is a set of Code that performs a specific action, often a Function that has been defined in the Requirements Document.
Negative Testing	Test Cases are divided into Positive Testing (testing transactions that are supposed to work), and Negative Testing (testing transactions that are not supposed to work).
	In Negative Testing, the edits or the logic of the Application should not allow the transaction to proceed, and should produce the appropriate Error or Warning Message.
	Positive Testing, Error Handling Testing
Operational Test	This is a type of structural test that validates that the Application is available as required and that it can be run in Production. This test goes beyond Functional Testing by ensuring that it works in the Environment under all of the anticipated Conditions. This is often included in the User Acceptance and Parallel Tests.
	Functional Test, User Acceptance Test, Parallel Test, Usability Test, Process Flow Test, Procedures Test,

Software Testing for Financial Firms   Page 310

	Simulated Work Environment, White Room Testing
Parallel Testing	Parallel Testing is usually managed by the End Users and is typically the last test before an Application goes live. If the new Application is replacing an existing one, both Applications are operated side by side, using the same input transactions, for a pre-determined period of time to ensure that the output from both Applications is the same, or that any differences are acceptable.
	Testing, Systems Development Life Cycle
Peer Review	A Peer Review is an independent review of a requirement, Design, Code Module, Test Plan, etc., by a person or group not directly involved with the activity.
Performance Test	Performance Testing is the testing that is conducted to evaluate the compliance of an Application or component with specified performance Requirements such as response time, batch-processing time, etc.
	Additional types of Performance Tests that can be conducted are Load Testing, Stress Testing, etc.
Pilot	A Pilot is a version of the intended Application that has very limited Functions. It is used to verify that the Applications Design meets the users' Requirements.
	Pilot Testing
Pilot Testing	Pilot Testing occurs when a new Application is installed and tested on a limited basis prior to full installation or deployment. It is used to verify that the Application's Design meets the users' Requirements and can also be helpful in making the decision to buy or build an Application.
Plan	A Plan is a logical list of Tasks, milestones, resource assignments and dependencies that when performed, will attain a goal.
	Goal, Task, Milestone, Project Plan
PLC	PLC is an abbreviation for Product Life Cycle or Project Life Cycle
	See Systems Development Life Cycle
Policy	A Policy is established by a firm as a rule that sets a general direction, or which establishes actions that cannot or should not be taken in certain circumstances. A Policy differs from a procedure in that the Policy sets the general Guideline, while a procedure defines how something should be done.
	Procedure
Positive Testing	Test Cases are divided into Positive Testing (testing transactions that are supposed to work), and Negative Testing (testing transactions that are not supported to work).
	In Positive Testing, the edits or the logic of the Application should correctly process the transaction.
	See Negative Testing
Predicted Results	See Expected Results
Problem/Incident Report	See Issues Tracking Report
Procedure	A firm establishes a Procedure to specifically define how a Task should be performed. A group of procedures constitute a Process.
	Policy, Process, Task
Procedures Test	Procedures Testing is usually performed as a part of User Acceptance Testing or Parallel Test to validate that the users can actually use the Application in the manner that it was intended. This is sometimes called White Room Testing, Simulated Work Environment, or Model Office testing.
	Usability Test, Operational Test, Process Flow Test, Simulated Work Environment, White Room Testing, Model Office Test
Process	A Process is the sequence of activities that results in a desired result. Processes are defined by procedures and methodologies that are established based upon Policies and Standards.
Process Flow	This is a type of test that follows a transaction from input though output across as many Applications as have a role in completing that transaction.
Process Flow Testing	Process Flow Testing tests the Application's ability to correctly process a transaction from the time it enters the Application until it is finally posted and reported. In some cases this can occur immediately, and in other cases the process flow might extend over several Production Days, as in securities settlements. Some types of transaction flows are: • Check Processing • Securities Settlement • Funds Transfers • Issuance • Insurance Policy
	Usability Test, Operational Test, Procedures Test, Simulated Work Environment, White Room Testing, Model Office Test, Production Day
Product Life Cycle	See Systems Development Life Cycle

Production	Production involves using an Application to process live data that has a real impact on the business and the Clients.
Production Day	A Production Day uses Production data that is related to an actual day of Production (i.e., April 2, April 3, etc.). It may take more than one Test Day to run a production day.
Program	The term Program can be used in two ways: - A Program is a group of lines of Code that is designed to perform a specific Task. A group of Programs working together is an Application. - A Program is a series of related events that is intended to achieve a specific goal and to maintain that goal over a period of time. A Program has a start date, but usually does not have a designated end date. Project, Application, Module
Programming Phase	This is the phase of the SDLC where Code is Developed. During this phase the testing team usually installs or Develops testing tools and Test Cases. In many Projects some Coding continues throughout the testing cycle. Systems Development Life Cycle
Project	A Project is a series of related events that are intended to achieve a specific goal in a specific period of time. A Project has a start date and an end date. Program
Project Life Cycle	See Systems Development Life Cycle
Project Plan	A Project Plan is the list of activities that must be performed in order to accomplish a Project's objectives. Project Plans can take many forms, although almost all of them contain at least: - Tasks - People responsible for the completion of the Tasks - Start dates for the Tasks - End dates for the Tasks - Dependencies of the Task with other Tasks
Project Risk	There are five possible outcomes for any Project, and only one of them is good. All of the others are categories of Risk that must be managed during a Project. 1. Goes live on time, within Budget, and is a success 2. Project is late and/or over Budget 3. Cancelled during Development 4. Goes live, but fails in Production Incorrect Requirements, unstable or poor quality Code) 5. Goes live but is cancelled before reaching its payback
QA	Quality Assurance
QA Plan	See Test Plan
QC	Quality Control
Quality Assurance	Quality Assurance (QA) is an organization or activity that can be used to ensure that the overall quality of a product, Project, Application or Program is within the specification defined by the users. QA includes testing. The process of assuring quality involves measuring, monitoring and managing. The long-term goal of most QA methodologies is to ensure immediate adherence to policies and procedures and establish an environment that supports continuous process improvement.
Quality Control	Quality Control (QC) is the methodology by which the actual output/results of a product, Project, Application or Program is measured and compared to an applicable Standard and then action is taken when a negative difference is detected.
Recovery Test	Recovery Testing validates that an Application can properly recover from crashes, hardware failures, or other catastrophic problems. Recovery Tests are performed at the beginning of a Production Day, at various critical points throughout the day and at the end of the Production Day. The goal is to ensure that: - The data is available - Any transactions that were not processed can be identified - The Application can be restarted in a reasonable amount of time - The End Users can perform their routine Tasks after the Applications have been restarted Contingency Testing, Back-up Testing, Production Day
Regression Test	A Regression Test is a type of test that ensures that no unexpected Changes emerged in one part of an Application as a result of making Coding Changes to another part of the Application. This test involves testing a previously tested Program after a Change has been made to verify that new Defects have not been introduced or uncovered as a result of the Change that was made. Regression Testing can occur during Functional, Business Integration and User Acceptance Testing, or after an Application has gone live. Functional Testing, Business Integration Testing, User Acceptance Testing
Report	A Report is an organized presentation of data to answer specific management questions. The same basic Report can be Sorted in different ways to hi-light different aspects of the data. Sort, View, MIS

Requirements	See Functional Specification, Functional Requirement
Requirements Document	This Document is produced during the requirement definition phase of the SDLC. Systems Development Life Cycle
Requirements Phase	The Requirements Phase of the SDLC consists of the Development of the Business Requirements and Functional Requirements Documents. Systems Development Life Cycle
Resource	A Resource is any person, Application, hardware item, software tool, etc., that that can be utilized for a specific purpose.
Risk	See Project Risk
Rule-Set	
Sanity Test	Sanity Testing is an informal activity that can occur during Functional Testing, User Acceptance Testing or Parallel Testing, and its purpose is to make a judgment regarding whether or not the testing should proceed based upon the Defects that have been discovered. Functional Test, User Acceptance Test, Parallel Test
Security Test	Security Testing verifies how well the Application protects against unauthorized internal or external access, willful damage, mischief, etc. The amount and depth of testing required will depend on the Risk assessment of the impact of a security breach. Security testing usually occurs during Unit, System Integration and User Acceptance Testing. Unit Test, Systems Integration Test, User Acceptance Test
Simulated Work Environment	See Procedures Test, White Room Testing, Usability Test, Process Flow Test, Model Office Test
Sort	Reports can be Sorted based upon different criteria in order to present information to the Test Manager in a way that makes it easy to understand, and which helps to identify specific problems or opportunities.
Stakeholder	A Stakeholder is a person who has a vested interest in the success of a particular Project, Program or Application.
Standard	A Standard is a prescribed way of presenting information or performing a Task. There are a variety of Standards used in the securities industry, such as pricing, Messaging and numbering Standards.
Standards Organizations	Standards organizations such as ANSI and IEEE provide Guidelines for Planning, performing and documenting the testing effort (as well as other phases of the SDLC). Each organization may also have internal Standards that should be adhered to.
Static Testing	Static Testing involves testing Code without actually Executing it. Walkthrough
Storage Test	A Storage Test is performed to determine if an Application's Design will effectively utilize memory and storage space either in resident memory or on disk. This usually occurs during Systems Technical Testing.
Stress Test	Stress Testing can be: <ul><li>Inserting a variety of assumptions into an investment model to see what the worst-case outcome of a portfolio could be.</li><li>Inputting a large number of transactions into a computer Application to see if all of the Components (Application, hardware, telecommunications, people, etc.) can accommodate the peak volume within acceptable time frames. A Stress Test is usually performed after User Acceptance Testing, and usually before Parallel Testing.</li></ul>Numerous aspects of an Application are exercised during a Stress Test, including:<ul><li>Ability of a Screen to accept and edit a large number of transactions simultaneously</li><li>Ability of an online or batch program to generate reports within a specified timeframe</li><li>Ability to access the same Tables simultaneously from various Programs</li><li>Ability to send and receive a large number of Messages in a specified timeframe</li><li>Ability to perform the required logical calculations on a large number of transactions in a specified timeframe</li><li>Ability to read and write large amounts of data to the database</li></ul>The term Stress Testing is often used synonymously with Load Testing, Volume Testing or Performance Testing. Some ways to stress an Application are:<ul><li>High volumes of users performing concurrent activities</li><li>High volumes of Messages in and out of the systems</li><li>High volumes of on-line transactions</li><li>High volumes in batch processing</li><li>Large data bases with complex queries in narrow timeframes</li></ul>Storage Test, Systems Development Life Cycle, Load Test, Volume Test, Performance Test
Structured Testing	Most forms of Testing described in this book are Structured Tests. In a Structured Test, the tester enters a pre-determined test Condition and compares the results to pre-defined Expected Results.

	Expected Results
System	A System is one or more Applications that work together with their supporting hardware to meet a business need.
	Application
Systems Development Life Cycle	A Systems Development Life Cycle (SDLC) is a process that identifies the steps that are needed to define, Design, Develop, test and install a computer Application. While there are several generic SDLC processes available, most firms create a customized version for their own use, and most contain the following elements: • Concept/Analysis Phase • Requirements Phase • Design Phase • Development Phase • Testing Phase • Implementation Phase Firms without an SDLC generally experience much higher Development costs, longer Development cycles, higher Maintenance costs and higher Project Risk. Project Risk
Systems Integration Test	Systems Testing, also called Systems Integration Testing, is a fully integrated test of an entire Application and is usually conducted by the Developers before they deliver it to the Test Team for the next step in the testing process. Some Development organizations do not fully test the Application against the user Requirements Document and do not discover Defects until user testing. This is very inefficient and leads to organizational disputes between the Development organization and the user departments. Systems Development Life Cycle
Systems Test	See Systems Integration Test
Task	A Task is an action that someone must perform in order to achieve a defined result. Procedure
Test Bed	See Test Library
Test Case	A Document specifying the test approach for a software feature or combination or features and the inputs, Expected Results and Execution Conditions for the associated tests.
Test Cycle	Most tests for financial services require that the testing be performed over a number of different Production Days. Each of these Test Days may be a Test Cycle, or several Test Days could be grouped in a Test Cycle.
Test Library	A Test Library is a set of Test Transactions that can be used to test specific Applications and which can be saved for future use. If properly organized, the Test Bed contains the Business Scenarios, the Test Cases and the Test Procedures.
Test Log	The Test Log records the results of a test.
Test Phase	In general, a Test Phase corresponds to a Test Type and can include: • A partial Test Type • An individual Test Type • A combination of Test Types
Test Plan	The Test Plan is a detailed Document that describes the Scope, approach, resources, and schedule of intended testing activities. It identifies the features to be tested, the Test Types to perform, the testing Tasks, who will do each Task, and any Risks requiring contingency Planning. Ref IEEE Std 829.
Test Report	The Test Report consists of the Test identifier, the Expected Results, the Actual Results and the follow-up action being taken, if any.
Test Script Procedure	The Test Script Procedure is the process that has been defined to describe the sequential actions and Expected Results of a test session and which may be automated. A paper-based Test Procedure will also contain the Test Report. Test Report.
Test Suite	A Test Suite is a set of Test Cases along with the rules for entering them.
Test Type	There are several Test Types discussed in this book: Business Analysis • Business Requirements • Functional Requirements • Business Review Design • Technical Design Review • Data Design Review Coding • Unit Test

Software Testing for Financial Firms Page 314

	• Systems Integration Test • Systems Technical Test • Test Team Testing • Functional Test • Business Integration Test • Business Technical Test • Regression Test End User Testing • User Acceptance Testing • Conversion Test • Parallel Test
Testing	Testing a computer Application is usually done in several steps, as defined in a Systems Development Life Cycle. The purpose of the testing is to verify that an Application meets the Requirements found in the Requirements Document, and if not, to identify the Defects.
Testing Region	The Testing Region is a discrete place within a computer that is assigned to the testing team for their exclusive use, and is one of three primary regions that are used: • Development (managed by the Developers) • Testing (managed by the Testers) • Production (managed by the IT Department) Access to the testing region is restricted. Only selected people can move Code into the Testing Region and from the Testing Region to the Production Region.
Text-Based User Interface	A Terminal Based (or Text Based) Interface is the original method used by people to interact with a computer. The computer displays words or numbers as output, and a prompt identifies when the user must input a command by typing in words, letters or numbers. Graphical User Interface
Transaction Flow	See Process Flow
Transaction Flow Testing	See Process Flow Testing
Transaction Type	Most Applications categorize their Functions into Transaction Types, which are generally associated with specific input and retrieval Screens.
TUI	See Text-Based User Interface
UI	See User Interface
Unit Test	Unit Testing is one of the first steps in the testing process, and is usually conducted by a Developer to ensure that each unit of Code Functions properly when exercised independently. Testing, Systems Development Life Cycle
Usability Test	This is a process to determine how well the users can learn and use a product in a routine day-to-day environment. It will often be performed during User Acceptance Testing. Operational Test, Process Flow Test, Procedures Test, Simulated Work Environment, White Room Testing
User Acceptance Test	One of the final tests in the Project, User Acceptance Testing is a very critical step since it is managed by the End Users to determine if the Application meets the terms of the Requirements Document. Acceptance Tests are usually defined and Executed by the End Users and contain a mix of normal and unusual situations that they have experienced in live processing. This test should consist of a series of predetermined tests, with defined Expected Results, that will validate the Function of the Application and ensure that the users can work with the Application as it has been Designed. Frequently, User Acceptance Testing also is used to achieve the following: • User Training on the new Application • Documentation Review • Workflow Validation Acceptance Criteria, Systems Development Life Cycle, Testing
User Interface	A User Interface is a Screen that allows the user of the Application to enter or retrieve data, and to initiate a Function. There are two types of UIs, Graphical User Interfaces and Text-Based Interfaces. Graphical User Interface, Text-Based User Interface
View	A View of data is a Sort of a Report that presents information in a specific way to the user of the Report. Sort
Volume Test	The Volume Test is a Stress Test that determines the largest number of users or transactions that a Program can correctly process in a specific period of time. Stress Test
Walkthrough	A Walkthrough is a manual process that can be performed at the Requirements, Design or Coding stage of the SDLC to determine if the Requirements, Design and/or Code will meet the users' needs.

	This usually involves knowledgeable people examining Requirements or Code to see if it seems correct to them, based upon their experience. While this is not a highly rigorous form of testing, it is very cost effective since errors found during this process are the least costly to correct.
White Box Testing	In order to conduct White Box Testing, the Tester must understand the internal workings of an Application. The White Box approach is essentially a Unit Test and Systems Integration method (and is sometimes used in other levels of testing) and is almost always performed by technical staff and can include: • Testing of branches and decisions in Code • Tracking of a logic path in a Program • Validating that a formula is calculating correctly
White Room Testing	See Operational Test, Process Flow Test, Procedures Test, Simulated Work Environment, Model Office Testing

Appendix VIII - Index

Acceptance Criteria 280, 281, 306, 314
Acceptance Test...... 27, 234, 238, 244, 248, 250, 251, 252, 254, 255, 259, 281, 306, 314
Ad Hoc Testing... 306
Application 1, 2, 3, 5, 6, 9, 10, 11, 12, 13, 14, 15, 16, 17, 19, 20, 21, 22, 23, 24, 25, 26, 27, 28, 29, 30, 31, 33, 34, 35, 36, 37, 38, 39, 40, 41, 42, 43, 44, 45, 46, 47, 48, 49, 51, 52, 53, 54, 61, 62, 63, 64, 65, 66, 68, 70, 71, 72, 73, 74, 76, 77, 79, 80, 81, 82, 84, 86, 87, 88, 89, 91, 97, 98, 99, 111, 112, 113, 114, 115, 118, 119, 120, 121, 123, 124, 125, 129, 131, 133, 135, 138, 139, 142, 147, 149, 150, 151, 152, 153, 171, 172, 179, 183, 184, 192, 194, 195, 209, 212, 213, 215, 216, 217, 218, 219, 220, 223, 225, 226, 227, 228, 232, 233, 235, 237, 238, 239, 241, 242, 243, 246, 247, 249, 251, 252, 253, 254, 255, 256, 257, 259, 262, 264, 265, 268, 269, 270, 271, 273, 279, 280, 281, 287, 290, 293, 294, 295, 296, 297, 298, 299, 300, 305, 306, 307, 308, 309, 310, 311, 312, 313, 314, 315
ASQ... 306
Audit Test... 306
Automated Software Quality 306
Automated Testing 211, 212, 213, 214, 306
Backup and Recovery Testing 306
Back-up Testing.................................... 306, 311
Black Box Test 21, 25, 61, 209, 221, 306
Boundary Scenario 306
Bug 43, 45, 64, 87, 213, 306
Business Integration Test......... 3, 24, 25, 26, 27, 31, 49, 65, 74, 76, 79, 85, 86, 103, 113, 155, 234, 238, 244, 248, 250, 251, 252, 254, 255, 259, 306, 308, 309, 311, 313
Business Scenario 15, 17, 25, 27, 55, 60, 74, 75, 77, 89, 95, 98, 111, 112, 113, 114, 116, 128, 129, 138, 184, 205, 302, 305, 306, 309, 313
Calendar Day 80, 83, 84, 88, 306, 309
Capture/Replay Tool.................................... 306
CAST... 306
Changes Only Testing........................... 26, 306
Checkpoint... 306, 309
Code.. ..... 1, 2, 3, 4, 5, 11, 12, 15, 16, 20, 21, 22, 23, 24, 25, 26, 33, 35, 37, 38, 39, 41, 42, 43, 45, 46, 49, 51, 52, 54, 61, 62, 63, 66, 72, 79, 96, 102, 120, 125, 139, 140, 156, 162, 164, 165, 166, 167, 169, 170, 171, 172, 174, 175, 176, 177, 178, 180, 182, 196, 198, 201, 209, 210, 211, 213, 214, 216, 221, 234, 235, 236, 237, 238, 239, 242, 245, 248, 249, 250, 252, 254, 255, 257, 259, 261, 280, 281, 298, 299, 301, 303, 305, 306, 307, 309, 310, 311, 312, 314, 315
Code Complete 307
Code Coverage 63, 307
Code Coverage Test.............................. 23, 307
Code Freeze 307
Code-Set 39, 49, 307

Coding Phase 311
Comparison Test 29, 31, 307
Compatibility Test......................... 23, 307, 309
Compliance Testing 307
Component 73, 77, 129, 130, 257, 307
Computer Aided Software Testing........ 306, 307
Concept Phase... 307
Conformance Test 23, 307
Contingency Test 23, 306, 307, 308, 311
Conversion Test 3, 13, 26, 29, 31, 46, 47, 48, 79, 109, 307, 313
Coverage Analysis 99, 307
Critical Success Factors 307
Data Validation 307
Debug... 242, 307
Defect..... 3, 4, 14, 21, 25, 43, 45, 46, 49, 54, 62, 82, 86, 87, 89, 115, 120, 123, 125, 135, 141, 144, 148, 156, 162, 170, 173, 175, 178, 196, 213, 216, 224, 239, 241, 246, 247, 249, 254, 257, 265, 280, 281, 301, 302, 306, 307, 309
Definition Phase.. 307
Dependency..69, 70, 71, 77, 78, 94, 95, 97, 106, 302, 303, 307
Dependency Project ... .69, 70, 71, 77, 78, 94, 95, 97, 106, 302, 303, 307
Design Phase........ 4, 12, 33, 39, 51, 66, 73, 307, 313
Desk Checking....................................... 20, 308
Disaster Recovery Test.......................... 23, 308
Documentation Review 28, 308, 314
Documentation Testing 308
Dynamic Testing... 308
End-to-End Testing 25, 308
Entry Criteria 308
Error Handling Tests................................... 308
Exhaustive Testing...................................... 308
Exit Criteria...................... 61, 79, 301, 303, 308
Expected Results ... 13, 14, 55, 62, 89, 112, 115, 116, 117, 118, 119, 122, 128, 165, 205, 209, 211, 215, 217, 226, 263, 269, 270, 305, 307, 308, 310, 312, 313, 314
Exploratory Testing..................................... 308
Feature 279, 308
Finding 4, 283, 308
Floor... iii
Functional Requirement .. 2, 3, 5, 11, 12, 16, 19, 20, 23, 33, 35, 38, 40, 43, 45, 47, 54, 61, 62, 65, 71, 76, 98, 117, 129, 132, 143, 149, 164, 169, 184, 205, 209, 226, 306, 308, 312, 313
Functional Specification.... 21, 40, 61, 147, 308, 312
Functional Test..... 3, 22, 24, 25, 29, 31, 44, 49, 62, 65, 74, 76, 79, 80, 85, 87, 96, 97, 98, 100, 101, 103, 109, 110, 111, 113, 119, 121, 129, 132, 133, 134, 135, 136, 142, 143, 144, 147, 152, 155, 158, 204, 233, 236, 256, 301, 305, 306, 308, 309, 311, 312, 313
Functional Testing..... ...22, 24, 25, 44, 87, 306, 308, 309, 312
Goal 308, 310

Graphical User Interface 131, 166, 279, 308, 314
GUI 98, 131, 132, 142, 143, 148, 150, 173, 192, 216, 217, 221, 234, 235, 240, 241, 245, 247, 249, 250, 251, 252, 256, 259, 279, 300, 303, 305, 308
Implementation Test 309
Incremental Integration Test 22, 308, 313
Information Technology 286, 309
Inspection 61, 215, 309
Installation Test 309
Integration Testing..... 12, 21, 22, 25, 65, 71, 72, 234, 238, 244, 248, 250, 251, 252, 254, 255, 259, 308, 309, 313
Interface Testing... 309
Interoperability Test....................... 23, 307, 309
Issue....... ..45, 120, 123, 124, 125, 126, 127, 128, 135, 168, 169, 171, 172, 173, 174, 181, 185, 190, 194, 196, 198, 224, 225, 280, 281, 309
Issues Tracking Log....................................... 309
Issues Tracking Report 45, 55, 60, 122, 123, 124, 125, 128, 173, 201, 203, 227, 263, 309, 310
IT 4, 5, 56, 88, 101, 109, 157, 201, 243, 261, 264, 265, 282, 283, 286, 287, 289, 290, 293, 309, 314, 321
Liability ii
Load Test...............................295, 309, 310, 312
Logical Day 80, 309
Management Information System 153, 309
Metrics.. 53, 309
Milestone.............. 107, 159, 306, 307, 309, 310
MIS....... ..54, 106, 153, 154, 156, 165, 167, 168, 172, 173, 174, 175, 178, 180, 181, 184, 185, 186, 187, 190, 192, 195, 198, 205, 206, 207, 211, 212, 301, 305, 309, 311
Model Office Testing..................... 28, 309, 315
Module 2, 3, 21, 38, 44, 133, 134, 135, 136, 141, 144, 150, 152, 233, 247, 297, 303, 306, 309, 310, 311
Negative Testing...................129, 308, 309, 310
Operational Test 28, 309, 310, 314, 315
Parallel Testing..................17, 86, 309, 310, 312
Peer Review.. 310
Performance Test....46, 155, 213, 310, 312, 314
Pilot 30, 32, 310
Pilot Testing 30, 32, 310
Plan...... 6, 12, 13, 14, 30, 33, 41, 49, 51, 52, 54, 60, 61, 66, 67, 68, 69, 70, 71, 72, 73, 77, 78, 79, 80, 83, 84, 86, 88, 89, 90, 92, 93, 94, 95, 96, 97, 99, 100, 101, 102, 103, 105, 106, 107, 108, 109, 110, 113, 117, 121, 129, 133, 136, 148, 154, 155, 156, 158, 160, 165, 177, 182, 183, 184, 186, 187, 188, 191, 193, 195, 198, 199, 200, 202, 203, 204, 205, 207, 218, 225, 226, 228, 251, 258, 262, 263, 303, 304, 306, 309, 310, 311, 313
PLC 1, 310
Policy...72, 116, 141, 310
Positive Testing 129, 309, 310
Predicted Results 55, 204, 205, 210, 310
Problem/Incident Report.............................. 310

Procedure 60, 100, 101, 112, 114, 115, 116, 118, 119, 120, 122, 125, 128, 130, 157, 158, 179, 180, 183, 185, 239, 247, 251, 253, 279, 297, 308, 310, 313
Procedures Test 28, 309, 310, 312, 314, 315
Process 6, 9, 28, 29, 43, 52, 58, 59, 61, 62, 76, 92, 96, 97, 100, 101, 111, 121, 129, 154, 157, 210, 218, 220, 225, 232, 234, 240, 242, 246, 249, 250, 252, 256, 257, 258, 260, 262, 263, 264, 267, 301, 304, 309, 310, 312, 314, 315, 320
Process Flow ...28, 210, 309, 310, 312, 314, 315
Process Flow Testing 28, 310, 314
Product Life Cycle 1, 310
Production 1, 2, 3, 4, 5, 10, 15, 17, 22, 25, 26, 27, 28, 29, 37, 38, 46, 75, 78, 80, 84, 86, 88, 89, 98, 99, 101, 109, 111, 112, 113, 119, 121, 124, 126, 127, 136, 164, 171, 176, 177, 186, 195, 209, 213, 218, 223, 224, 225, 226, 241, 243, 244, 258, 268, 280, 281, 304, 305, 306, 307, 309, 310, 311, 313, 314
Production Day 80, 309, 310, 311, 313
Program.........2, 3, 21, 25, 29, 47, 53, 57, 63, 92, 124, 164, 209, 211, 213, 242, 304, 306, 311, 312, 314, 315
Programming Phase 4, 11, 12, 311
Project.......1, 2, 3, 4, 6, 9, 10, 11, 12, 13, 14, 15, 16, 17, 18, 19, 24, 30, 33, 34, 35, 36, 39, 41, 46, 49, 51, 52, 53, 54, 57, 58, 61, 64, 65, 66, 67, 68, 69, 70, 71, 72, 73, 77, 78, 79, 82, 83, 84, 85, 86, 88, 90, 91, 92, 93, 94, 95, 96, 97, 98, 99, 100, 101, 102, 103, 105, 107, 108, 109, 110, 111, 114, 123, 124, 126, 127, 131, 135, 136, 141, 153, 154, 155, 156, 157, 159, 160, 161, 162, 163, 164, 165, 168, 169, 170, 171, 172, 173, 174, 175, 177, 178, 179, 181, 185, 186, 187, 190, 191, 192, 193, 194, 195, 196, 197, 198, 199, 200, 201, 202, 203, 204, 205, 206, 207, 210, 212, 213, 214, 223, 224, 226, 227, 228, 229, 230, 231, 232, 253, 254, 261, 262, 263, 265, 267, 269, 270, 280, 281, 282, 284, 294, 302, 304, 307, 309, 310, 311, 312, 313, 314, 320, 321
Project Life Cycle 1, 310, 311
Project Plan. 6, 10, 13, 41, 52, 54, 69, 70, 83, 86, 90, 93, 94, 95, 97, 100, 102, 103, 105, 107, 108, 109, 110, 153, 154, 155, 156, 159, 160, 178, 181, 202, 203, 205, 207, 263, 265, 269, 302, 304, 310, 311
Project Risk..................................... 311, 312, 313
QA 5, 58, 63, 237, 253, 257, 282, 286, 287, 288, 289, 304, 311
QA Plan..311
QC 311
Quality Assurance....... ...5, 6, 57, 69, 70, 92, 94, 106, 164, 165, 210, 262, 286, 304, 311
Quality Control 311, 320
Recovery Test 23, 306, 307, 311

Regression Test 3, 24, 26, 31, 44, 46, 62, 63, 65, 74, 79, 86, 91, 113, 155, 156, 213, 223,

227, 243, 246, 247, 253, 257, 263, 306, 311, 313
Report.... 21, 45, 60, 76, 77, 78, 92, 97, 98, 101, 107, 108, 116, 118, 125, 132, 138, 143, 144, 145, 146, 147, 156, 157, 158, 159, 166, 167, 168, 169, 170, 171, 172, 173, 174, 175, 176, 177, 178, 182, 183, 184, 185, 187, 188, 189, 190, 191, 192, 197, 198, 202, 204, 206, 228, 229, 241, 243, 303, 305, 309, 311, 314
Requirements....... 1, 2, 3, 4, 5, 6, 10, 11, 12, 14, 15, 16, 19, 20, 21, 25, 35, 37, 40, 41, 42, 45, 47, 51, 52, 54, 58, 61, 62, 65, 66, 68, 70, 71, 72, 79, 82, 90, 94, 95, 102, 103, 106, 109, 117, 118, 134, 147, 148, 150, 151, 164, 173, 174, 176, 177, 182, 184, 185, 194, 195, 205, 207, 210, 212, 214, 218, 220, 226, 239, 247, 253, 264, 293, 301, 303, 304, 307, 308, 309, 310, 311, 312, 313, 314
Requirements Document....2, 16, 20, 35, 61, 62, 134, 147, 207, 307, 308, 309, 312, 313, 314
Requirements Phase........ 11, 307, 308, 312, 313
Resource 63, 66, 69, 70, 82, 89, 90, 91, 94, 103, 106, 107, 108, 109, 159, 191, 206, 296, 312
Risk.......... 5, 28, 52, 53, 69, 84, 86, 92, 98, 106, 141, 153, 171, 175, 177, 180, 194, 195, 196, 197, 198, 202, 204, 205, 206, 207, 264, 301, 304, 311, 312, 313
Rule-Set 35, 38, 39, 41, 96, 307, 312
Sanity Test.. 312
Sanity Testing... 312
Securities Operations Forum........................... iii
Security Test... 23, 312
Simulated Work Environment ..... .28, 309, 310, 312, 314, 315
Sort 135, 167, 168, 170, 172, 184, 188, 192, 311, 312, 314
Specifications 40, 45, 205, 264
Stakeholder.. 312
Standard..... .5, 55, 57, 58, 59, 60, 62, 67, 69, 70, 72, 107, 136, 184, 198, 235, 254, 257, 296, 307, 308, 309, 311, 312
Standards 52, 55, 56, 57, 67, 312
Standards Organizations 55, 312
Static Testing....................................... 308, 312
Storage Test................................... 23, 309, 312
Street .... ii, iii
Stress Test...... 12, 23, 24, 48, 65, 120, 211, 220, 233, 247, 291, 293, 309, 310, 312, 314
Structured Testing............................... 306, 312
System........ 2, 12, 17, 21, 22, 39, 40, 41, 42, 43, 44, 46, 48, 66, 71, 72, 78, 101, 133, 134, 147, 156, 157, 158, 162, 184, 196, 218, 234, 238, 244, 248, 250, 251, 252, 254, 255, 259, 280, 287, 290, 303, 309, 312, 313, 320
Systems Development Life Cycle... ...1, 2, 6, 9, 20, 51, 307, 308, 310, 311, 312, 313, 314
Systems Integration Test....3, 11, 20, 22, 31, 65, 74, 75, 79, 103, 155, 156, 281, 308, 309, 312, 313
Systems Test.................................. 72, 303, 313

Task ...2, 51, 84, 90, 93, 94, 101, 103, 107, 109, 110, 123, 136, 153, 154, 157, 159, 165, 172, 202, 264, 301, 306, 308, 309, 310, 311, 312, 313
Test Bed.. 313
Test Case....6, 14, 26, 36, 37, 55, 60, 61, 62, 63, 67, 75, 80, 84, 89, 96, 99, 100, 101, 103, 109, 111, 112, 113, 114, 116, 117, 118, 120, 125, 128, 130, 131, 132, 133, 134, 135, 136, 137, 138, 139, 140, 141, 142, 144, 145, 146, 147, 148, 149, 150, 151, 156, 157, 158, 159, 161, 162, 163, 164, 165, 169, 175, 176, 178, 179, 180, 183, 184, 185, 200, 204, 205, 206, 211, 213, 214, 217, 223, 227, 242, 247, 253, 257, 258, 263, 264, 293, 301, 302, 303, 305, 307, 308, 309, 310, 311, 313
Test Cycle 80, 81, 82, 86, 87, 116, 161, 162, 163, 164, 303, 313
Test Library......... 26, 36, 62, 120, 183, 214, 227, 269, 313
Test Log ...55, 60, 112, 115, 116, 119, 120, 122, 128, 313
Test Phase... .....4, 38, 66, 68, 69, 72, 75, 77, 78, 79, 81, 82, 83, 85, 86, 90, 91, 94, 95, 96, 99, 100, 101, 102, 103, 106, 108, 110, 111, 155, 156, 161, 162, 163, 164, 188, 189, 190, 191, 192, 203, 204, 206, 207, 209, 225, 228, 303, 306, 313
Test Plan 1, 6, 10, 13, 14, 27, 30, 33, 35, 39, 40, 41, 42, 43, 49, 51, 54, 55, 60, 61, 66, 67, 68, 69, 70, 71, 72, 73, 74, 75, 76, 77, 78, 79, 80, 81, 86, 87, 88, 89, 90, 92, 93, 94, 95, 96, 97, 98, 99, 100, 101, 102, 103, 105, 106, 108, 109, 110, 111, 112, 113, 116, 117, 120, 122, 129, 130, 153, 202, 203, 205, 212, 226, 232, 247, 251, 253, 256, 258, 263, 268, 269, 301, 302, 303, 304, 305, 306, 310, 311, 313
Test Report........................... 168, 178, 302, 313
Test Script......... 60, 73, 100, 101, 109, 117, 118, 119, 120, 179, 183, 313
Test Script Procedure 313
Test Suite 217, 246, 313
Test Type. 3, 6, 10, 11, 19, 20, 21, 22, 24, 26, 29, 30, 31, 33, 34, 38, 39, 41, 44, 45, 46, 48, 49, 55, 65, 67, 68, 71, 72, 74, 75, 76, 77, 78, 79, 90, 95, 99, 110, 113, 116, 121, 155, 215, 303, 309, 313
Testing... ...i, 2, 3, 4, 5, 6, 9, 10, 11, 12, 14, 15, 16, 17, 19, 20, 21, 22, 23, 24, 25, 26, 27, 29, 30, 34, 35, 36, 37, 38, 39, 40, 41, 42, 44, 46, 48, 49, 51, 52, 54, 57, 60, 62, 65, 66, 68, 71, 72, 73, 75, 76, 77, 78, 83, 86, 88, 89, 91, 92, 93, 95, 96, 97, 100, 105, 108, 109, 111, 113, 124, 126, 128, 129, 131, 137, 139, 141, 143, 144, 148, 155, 159, 161, 165, 173, 174, 176, 177, 179, 188, 190, 196, 202, 203, 204, 209, 210, 211, 212, 213, 214, 218, 220, 221, 226, 230, 233, 236, 238, 242, 248, 250, 255, 256, 257, 258, 262, 265, 267, 279, 283, 285, 286, 287, 289, 290, 294, 299, 302, 303, 304, 306, 307, 308, 309, 310, 312, 313, 314, 315

Testing Region...................................... 307, 314
Text Based Interface..................................... 314
Text-based User Interface.................... 131, 314
The Summit Group .... ii, iii
Transaction Flow 314
Transaction Flow Testing 314
Transaction Type ... 45, 46, 96, 98, 99, 112, 118, 122, 169, 176, 182, 225, 314
TUI 314
UI ...62, 131, 133, 134, 142, 146, 147, 148, 173, 314
Unit Test.... 3, 11, 12, 20, 21, 22, 25, 30, 65, 74, 76, 78, 79, 86, 87, 91, 103, 109, 138, 155, 156, 161, 162, 163, 164, 189, 209, 214, 281, 307, 308, 309, 312, 313, 314, 315
Usability Test............ 28, 63, 309, 310, 312, 314

Usability Testing 63, 314
User Acceptance Test.. 3, 17, 25, 26, 27, 28, 31, 66, 74, 79, 85, 113, 155, 214, 285, 306, 308, 309, 310, 311, 312, 313, 314
User Interface........25, 42, 43, 45, 62, 74, 76, 77, 97, 111, 115, 129, 130, 131, 133, 134, 135, 136, 140, 141, 142, 143, 145, 149, 151, 152, 166, 279, 284, 308, 314
View..... 107, 108, 133, 167, 168, 171, 176, 182, 184, 187, 188, 191, 192, 193, 202, 203, 253, 311, 314
Volume Test.......................... 211, 309, 312, 314
Walkthrough 20, 22, 309, 312, 314
White Box Testing 21, 25, 209, 221, 315
White Room Testing 28, 309, 310, 312, 314, 315

Appendix IX - Author Biographies

Hal McIntyre

Hal has over 30 years of management experience, including 16 years at Citibank. He founded The Summit Group in 1991. Prior to establishing The Summit Group, Hal managed numerous banking and securities functions for Citibank, including marketing, operations, and technology. During his last four years at Citibank, he managed an Applications Development Division of over 300 people who supported Citibank's worldwide securities processing applications.

While at Citibank, he worked for four years in Zurich as the head of Operations and Technology for Citibank's Swiss Investment and Private Banks, and was the Chief Administrative Officer for Citicorp Investment Bank.

Hal is a Managing Partner of The Summit Group, which, with over fifty employees, is active in consulting to major banks, brokers, investment managers, and other financial institutions, and provides comprehensive services in Management Consulting, Market Research, and Systems Integration.

Securities Operations Forum, a subsidiary of The Summit Group, is a leading supplier of information and training to the securities industry. Securities Operations Forum publishes two newsletters for over 4,000 readers worldwide: *Securities Operations Letter* focuses on the US industry, and *Global Custody News* concentrates on international securities processing. Securities Operations Forum also conducts open enrollment and customized training for over 2,500 people each year throughout the US on processing topics such as Derivatives, Corporate Actions, etc., and annually presents several major conferences on key securities industry issues.

A former Air Force officer, Hal has a Bachelor of Arts in Industrial Psychology from Miami University and an MBA from Southern Illinois University. He has taught graduate and undergraduate courses at Fairleigh Dickinson University, and has been a guest lecturer at the University of Massachusetts. He is frequently quoted in publications such as American Banker, Trends and Institutional Investor, is a regular speaker at securities industry conferences around the world, and is the author of How the US Securities Industry Works and the *Securities Operations Glossary*.

hal@tsgc.com

Deborah Fortuna

Deborah has over 20 years of management experience, primarily in the Financial Services Industry in the areas of Trust, Custody, Cash Management and Commercial Lending. She is currently providing management consulting services in support of Project Management, Quality Control, Business Process Re-engineering, Training and Project Office Implementation. Previously, Deborah was a Director at Deutsche Bank where, for over 8 years, she oversaw numerous large-scale systems implementations for the Retirement Services Division. Her emphasis was on the management of various phases of the System Development Life Cycle including system selections, business requirements, testing, business process re-engineering, conversions, training, and client rollouts for several multi-currency Trust and Fund Accounting Systems.

Prior to her tenure at Deutsche Bank, Deborah provided similar development and implementation support services to a variety of businesses, across multiple industries and system platforms. As a firm believer in the value of a structured and focused testing process, she has spent a substantial portion of her professional career helping organizations evaluate their testing needs and establishing the optimal organizational structure, test methodology and management processes to ensure the delivery of timely, high-quality, full-functioning products.

Deborah holds a Bachelor of Science degree from Pace University in Professional Computer Studies from Pace University in New York City.

deb.fortuna@tsgc.com

Megan Johnson

Megan has over 15 years experience in a wide range of software-related activities, including software design, testing and implementation of financial and non-financial business applications. She has worked for Key Financial Systems and NCS on Trust, Real Estate and Oil and Gas Accounting systems in support of design, testing and Conversion activities.

In addition, Megan consulted on major Projects with First Fidelity, Merchants Bank, UPS, Bankers Trust and Deutsche Bank. Megan is a recognized expert in all areas of testing from planning through to implementation, and has spent most of her consulting assignments in senior testing roles. She has participated in many high-exposure projects, including new product development and the Conversion of a mission critical application to a vendor package. In addition to her testing expertise, she has experience as a Business Analyst and as a Project Manager on several complex financial application projects.

Prior to entering the IT arena, Megan worked in the Trust Accounting and Securities Lending areas for Prudential Insurance Co.

Megan has a degree in Business Administration and Accounting from The College of New Rochelle in New Rochelle, NY.

megan.johnson@tsgc.com